The Learning Works

EARTH BOOK FOR KIDS

ACTIVITIES TO HELP HEAL THE ENVIRONMENT

Written by Linda Schwartz
Illustrated by Beverly Armstrong

The Learning Works

Cover Design & Illustration:
Beverly Armstrong

Cover Layout & Graphic Production:
Margy Brown

Text Design & Editorial Production:
Sherri M. Butterfield

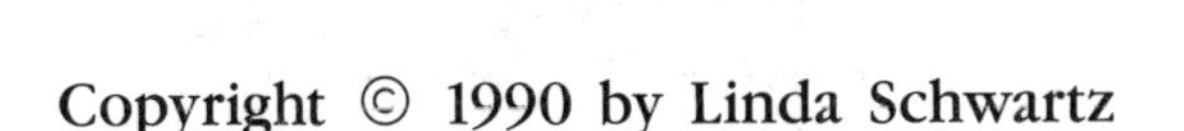

The Learning Works, Inc.

P.O. Box 6187

Santa Barbara, California 93160

Library of Congress Catalog Number: 90-091737

ISBN 0-88160-195-0

Printed in the United States of America. Current Printing (last digit): 10 9 8 7 6 5 4 3 2 1

Dedication

This book is dedicated
to
Stephen Howard and Michael Aaron
and to
children everywhere
who hold the future in their hands and
who care deeply enough
to take the first step
toward making the world a better place
in which to live.

Acknowledgments

Special thanks to William Schwartz, Bobbe Dartanner, Gene Mullins, Christina Lange, Madeline Sargent, Walter Larrabee, and Ray Rolland; to Brett Abbott, Chloe Coventry, Julie Dobie, Brian Haeberle, Sarah Kitson, Michael Schwartz, and Steven West for letters and poems; to Harry Hui and Peggy Bryant; to Tina Pickard for typesetting; to Margy Brown for cover layout and graphic production; to Sherri Butterfield—editor extraordinaire; to illustrator Beverly Armstrong in appreciation of her unique talent and infinite creativity; to each environmental organization that generously provided me with information; to all of the concerned individuals who graciously shared their expertise; to my family and especially to my husband Stan for his encouragement, support, and love.

—Linda Schwartz

Contents

Fact

One-fifth of the energy consumed in the United States is used for lighting.

Fact

For each ton of paper that is recycled, 24,000 gallons of water are saved.

Fact

The Environmental Protection Agency (EPA) wants Americans to be recycling 25 percent of their total trash production by 1992.

Contents
(continued)

EARTHWORDS

Man shapes himself through decisions that shape his environment.
—René Dubos

Contents (continued)

Fact

Less than one percent of the billions of pounds of pesticides used each year in the United States actually reaches a pest!

Contents
(continued)

PLANT & ANIMAL HABITATS 105–130

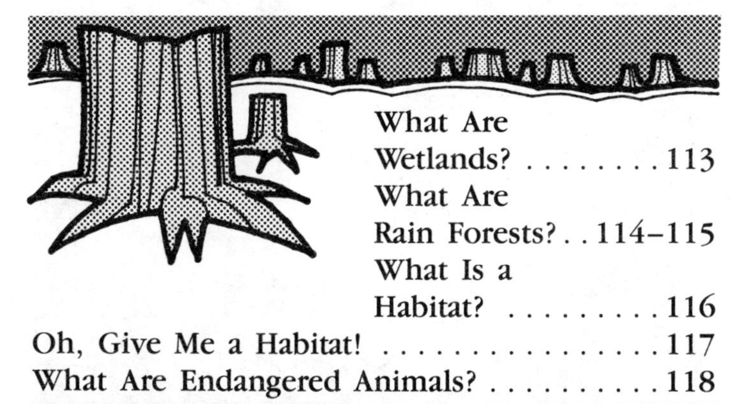

EARTHWORDS

Let us permit Nature to have her way; she understands her business better than we do.

—Michel de Montaigne

Contents
(continued)

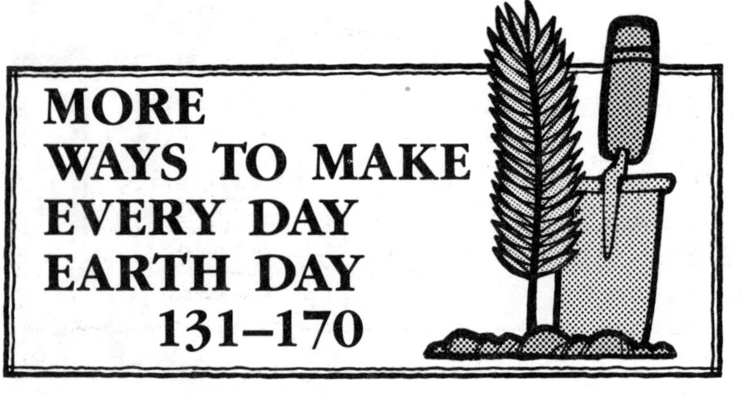

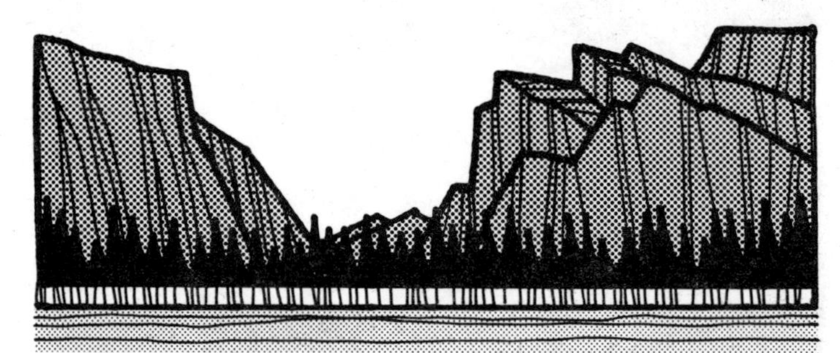

Contents
(continued)

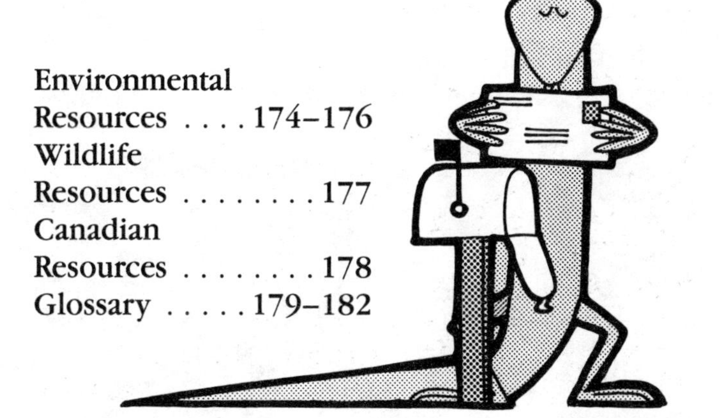

EARTHWORDS

There is no force greater than that of an idea whose time has come.

—Victor Hugo

A Note to Kids

Today's headlines tell of acid rain, dirty air, littered streets, oil spills, polluted beaches, power blackouts, the destruction of tropical rain forests, and the slaughter of endangered animals. The bad news is that the environment is in trouble. The good news is that you can help.

Earth Book for Kids was written to show you some of the ways in which you can make a difference. The facts presented in this book will help you understand the problems. The arts-and-crafts ideas, experiments, recycling projects, research topics, and other activities will help you become involved and discover some solutions. The earthwords will introduce you to other people who have cared about the earth. They have appreciated its beauty and worried about its future. They have written to express their appreciation and to persuade others to share their concern.

When you read, use, and enjoy this book, you don't necessarily have to start at the beginning and work your way to the end. Instead, begin with what interests you. Work alone or with family members and friends. Do what you can to make a difference where you are. And don't ever become discouraged or stop. You don't need to heal the entire earth; you just need to care about it and help a small part of it get better.

ENERGY, RESOURCES, & RECYCLING

Facts in Brief

American households and businesses generate about 160 million tons of solid waste, or garbage, each year.

Every Sunday, more than 500,000 trees are used to produce the 80 percent of newspapers that are never recycled.

Half of the present 6,000 landfills in the United States will be filled to capacity by the year 2000.

More than 24 million tons of grass clippings, leaves, and other yard wastes are thrown away annually in the United States.

One of the best weapons you have in the ongoing battle against garbage is to buy products that have been recycled.

Facts in Brief
(continued)

The average person in the United States uses about 580 pounds of paper annually.

People in the United States produce enough trash to fill 63,000 garbage trucks each day!

Plastics make up about 7 percent of our garbage by weight but 25 to 30 percent by volume. When discarded outdoors or placed in landfills, plastics become a permanent part of the landscape because they take hundreds of years to decompose.

In the United States, more than 25 billion styrofoam cups are thrown away each year. If all of these cups were placed in a line, the line would be long enough to circle the earth 436 times!

What Is Recycling?

Recycling is processing and treating discarded materials so that they can be used again. Materials that are commonly recycled include aluminum, glass, and paper. Recycling helps the environment in three very important ways.

When we recycle, we save space. Things that would have been thrown away are kept and reused. Thus, fewer discards find their way to crowded city dumps and bulging landfills. Outdoor spaces can be left open instead of being filled to capacity with mountains of trash.

When we recycle, we save energy. Of course, some energy is needed for the recycling process—to melt aluminum, to crush glass, or to convert newsprint into clean paper that can be used again—but recycling requires less energy than making new products from raw materials.

When we recycle, we save natural resources. In the recycling process, old materials are made into new products so fewer raw materials are used. Also, some of the coal, natural gas, water, or wood that might have been used to produce energy for the manufacturing process is not needed.

Recycling saves space, energy, and resources—three things we can no longer afford to waste—and it helps to reduce air and water pollution.

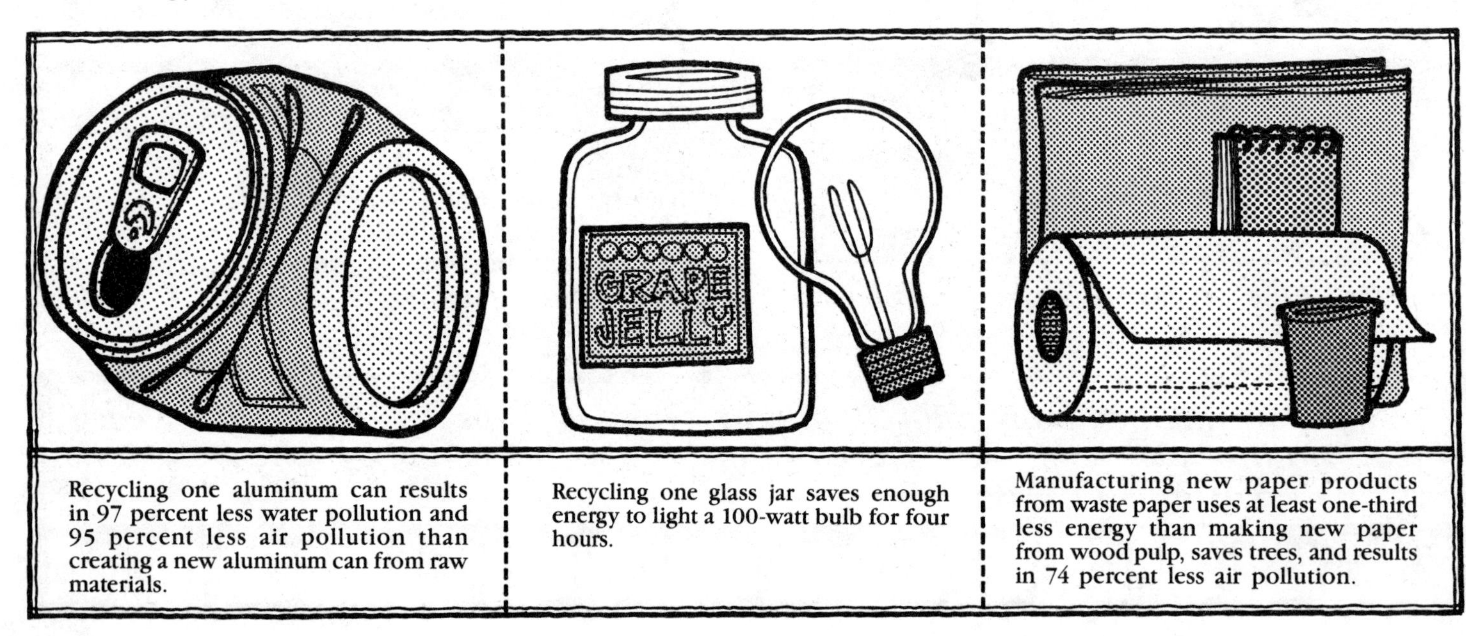

Recycling one aluminum can results in 97 percent less water pollution and 95 percent less air pollution than creating a new aluminum can from raw materials.

Recycling one glass jar saves enough energy to light a 100-watt bulb for four hours.

Manufacturing new paper products from waste paper uses at least one-third less energy than making new paper from wood pulp, saves trees, and results in 74 percent less air pollution.

Examples of Recycling

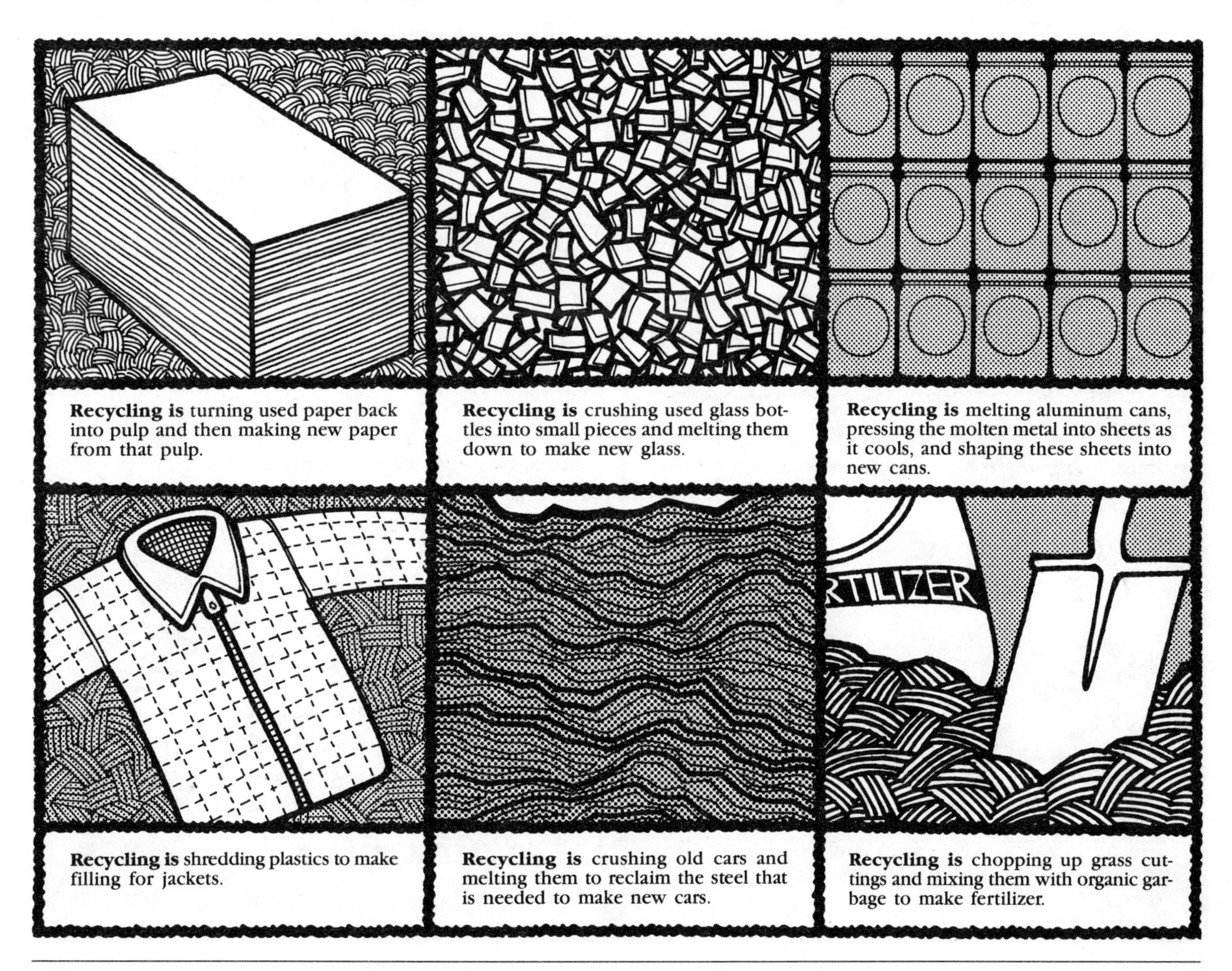

Steps in Recycling

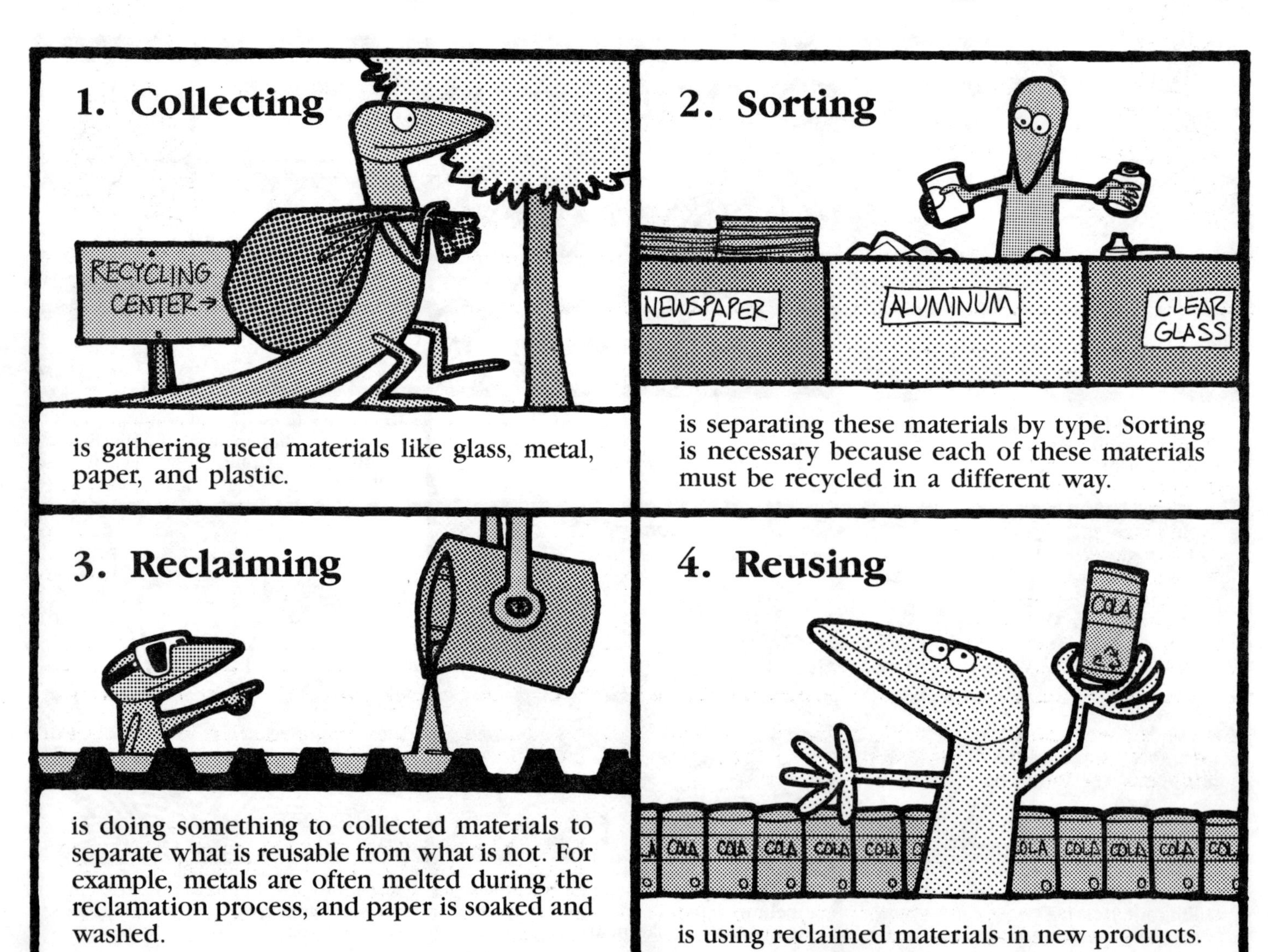

Recycling Survey

Conduct a survey to learn more about the recycling habits of your relatives, neighbors, and classmates. Use the sample questions listed below or make up your own. Summarize your results and share your findings with survey participants.

SAMPLE SURVEY QUESTIONS

	Yes	No	Do Not Use
1. Do you currently reuse and/or recycle any of the following items? (Answer *yes*, *no*, or *do not use*.)			
a. aluminum cans			
b. aluminum foil			
c. cardboard boxes			
d. computer paper			
e. glass bottles and jars			
f. magazines			
g. newspapers			
h. paper grocery bags			
i. plastic bottles and tubs			
j. plastic grocery bags			
k. polystyrene foam food containers and trays			
l. telephone books			

2. If you make a conscious effort to reuse and/or recycle, which ones of these statements reflect the most important reasons you do so?

a. I want to save natural resources.
b. I want to save energy.
c. I want to prevent air and/or water pollution.
d. I want to save money.
e. I am concerned about the environment, and reusing/recycling is one way I can make a difference.

3. If you do not make a conscious effort to reuse and/or recycle, which ones of these statements reflect the reasons you do not?

a. I really haven't given it much thought.
b. I don't have enough space to collect recyclables.
c. I don't know where the nearest recycling center is.
d. It's too much trouble.
e. I know the environment is in trouble, but I really don't care.

Looking for Litter

Litter is any unneeded item that has been carelessly discarded instead of being disposed of properly. Litter takes away from the beauty of the environment and can cause health-related problems as well. How much litter can you find in your community? Look for litter in places like the ones listed below. Do not handle the litter you find. Instead, count the pieces you see and record the numbers on a chart like this one.

LITTER CHART

Types of Litter	Places to Look			
	Your Yard	Your Neighborhood	Park/Playground	Downtown
bottles				
boxes				
cans				
discarded furniture				
newspapers				
paper wrappers				
plastic items				
toys				

A Litter Poem

Write a poem about litter. Your poem can be humorous or serious, rhymed or unrhymed. Here are several examples to get you started.

LETTER POEMS

In a letter poem, each letter of the topic or title of the poem is used as the initial letter for one line of the poem. The lines in the letter poem need not rhyme.

Great mounds of trash
All heaped up high
Ruin the beauty of earth.
Be aware of recycling.
All of us count.
Get started today.
Earth is ours to care for.

COUPLETS

A couplet is two rhymed lines. Many longer poems consist of a series of couplets.

Litter, litter everywhere!
Pick it up and show you care.

BEACHES

Beaches are meant for swimming and fun,
For playing Frisbee and soaking up sun.
Beaches are perfect for buckets and balls,
For building castles with moats and with walls.
And it's fun to go barefoot, as everyone knows,
Letting moist sand squish between wiggly toes.
But don't leave behind litter to spoil the beach.
A trash can is often within easy reach.

EARTHWORDS

The three great elemental sounds in nature are the sound of rain, the sound of wind in a primeval wood, and the sound of outer ocean on a beach.

—Henry Beston
The Outermost House

What Does Nature Recycle?

Do this experiment to learn which materials will decompose naturally and which will not.

WHAT YOU NEED

- ☐ a clay flowerpot
- ☐ a small stone
- ☐ enough soil to fill the flowerpot
- ☐ assorted litter, including foil, food scraps, leaves, paper, a plastic bag, and some polystyrene foam
- ☐ a pair of scissors
- ☐ some water
- ☐ a glass pie plate
- ☐ four weeks to wait for results
- ☐ some newspapers
- ☐ a stick about 12 to 18 inches long
- ☐ a pair of rubber gloves (optional)

WHAT YOU DO

1. With the stone, cover the hole in the bottom of the flowerpot so that water will not drain out too rapidly.

2. Put soil in the pot until it is about one-third full.

3. Cut, crush, tear, or break the litter into quarter-sized bits and pieces.

4. Scatter the litter over the soil.

What Does Nature Recycle?
(continued)

5. Cover the litter with soil until the pot is almost full.

6. Sprinkle the soil with water until it is thoroughly dampened but not completely soaked.

7. Cover the flowerpot with the glass pie plate.

8. Place the flowerpot in a warm, dark place.

9. Check the soil in the pot regularly and add water as needed to keep it moist.

10. After four weeks, empty the contents of the flowerpot onto open sheets of newspaper.

11. Put on the gloves if you intend to handle the soil, or use your stick to spread the soil so you can see what has happened to the litter.

12. Carefully observe the litter. Which materials decomposed? Which did not? What does nature recycle? Read pages 24–25 to learn more about our overcrowded landfills and pages 43–45 to learn more about the special problems posed by plastics and polystyrene foam.

What Is a Landfill?

A **landfill** is an enormous hole where garbage is dumped. To create a landfill, public works or sanitation engineers find a natural pit where low-lying land is surrounded by hills. If necessary, they use earth-moving machines to deepen the bottom and build up the sides of the pit. Usually, they line the inside of the pit to prevent the garbage from contaminating groundwater.

Once the pit has been shaped and lined, garbage is trucked in and dumped. Tractors spread the garbage in the landfill and cover each layer of garbage with dirt.

About 80 percent of the garbage we produce ends up in landfills. Of the remaining 20 percent, 10 percent is **incinerated**, or burned. Only 10 percent is recycled.

There are about 6,000 landfills in the United States. By the year 2000, half of them will be filled to capacity. Then, what will we do with our growing mountains of garbage? Where will we put it all?

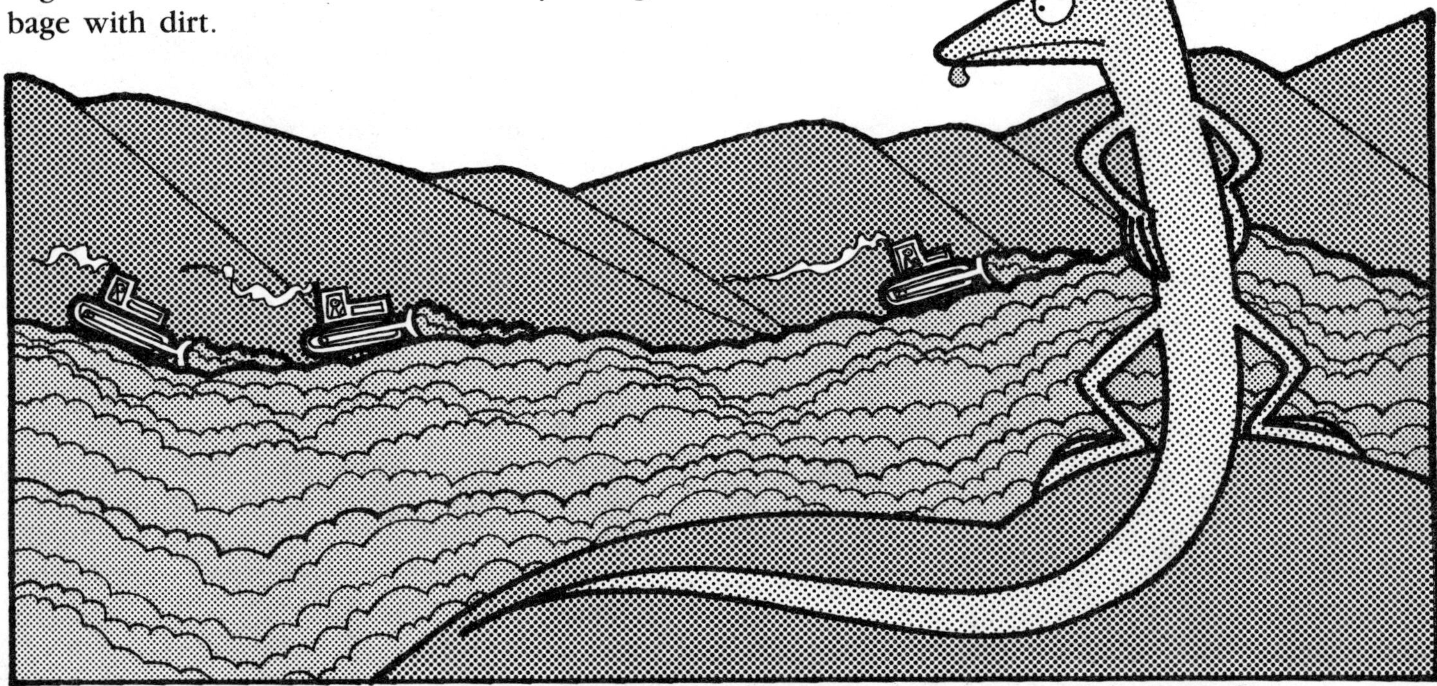

Do Something About Garbage

1. Do some research to learn where the garbage collected from your home eventually ends up. Is it trucked to a landfill? If so, where is that landfill and how close is it to being full?

2. Look up the telephone number for the Department of Sanitation or the Department of Public Works in your community. Call this number and ask how many tons of garbage are disposed of each week. If possible, find out how much garbage has been disposed of each year for the past five years. Is the amount of garbage produced by your community increasing or decreasing? Why?

3. Find the heading **Recycling** in the Yellow Pages of your telephone directory. Read the listings under this heading to locate the recycling centers that are operating in your community. Call or visit them to learn how they process goods for recycling. Then help the recycling effort by doing one of the following activities.

- **Start an educational campaign** to tell classmates and/or other community residents about the work of these recycling centers. Use fliers, posters, letters to the editor, and/or public service announcements in your campaign.
- **Sponsor a drive** to collect recyclable materials for one of these centers.
- **Create a display** to make people more aware of the acute landfill space shortage and to encourage them to join you in your recycling efforts. Use charts, graphs, drawings, photographs, recent magazine or newspaper articles, your paper trash sculpture (see page 26), and any statistics you can gather to make your point.

A Paper Trash Sculpture

Turn paper trash into a work of art.

1. Collect cardboard tubes (left from paper towel and toilet tissue rolls), egg cartons, junk mail, magazines, newspapers, paper cups, paper plates, pizza boxes, tissue and wrapping paper, and/or other paper trash.

2. Use staples, tape, and/or quick-drying glue to put these pieces of trash together into a sculpture.

3. Give your sculpture a finished look by mounting it on a cardboard base. A box lid or tablet back might be ideal.

4. Give your sculpture a title.

5. Share your sculpture with others by displaying it in your home or at school.

VARIATION. Incorporate metal or plastic trash into your paper sculpture or make a separate sculpture using these materials. In choosing your materials, consider metal bottle caps and pull tabs or plastic six-pack rings, margarine tubs, and lids.

Packaging Scavenger Hunt

How much aluminum, cardboard, cellophane, foil, paper, plastic, and polystyrene foam is used to package the foods you eat? The next time you are in the grocery store or supermarket, go on a packaging scavenger hunt. See how many foods you can find with no wrapping, with one wrapping, with two wrappings, and with three or more wrappings. List the foods you find on a chart like the one shown below.

FOODS WITH NO WRAPPING	FOODS WITH ONE WRAPPING
Example: apples 1.________ 2.________ 3.________	***Example: bread*** (in a paper or plastic bag) 1.________ 2.________ 3.________
FOODS WITH TWO WRAPPINGS	**THREE OR MORE WRAPPINGS**
Example: cereal (in a paper bag inside a cardboard box) 1.________ 2.________ 3.________	***Example: microwaveable meals*** (on a plastic tray with a foil lid inside a cardboard box) 1.________ 2.________ 3.________

Redesign a Package

The next time you are in a grocery store or supermarket, select one food product that has several layers of packaging. Redesign the packaging so that fewer layers and types of wrap are used. When possible, use materials that can be recycled. Make sure that your new packaging (1) provides plenty of space to identify the product, list the ingredients, and give nutritional information; (2) keeps the product fresh; (3) protects the product from damage; (4) prevents contamination; (5) encourages purchase; and (6) is both safe and convenient for members of the buying public to handle, store, and use. Draw and label a sketch of your new packaging or make a model of it.

EARTHWORDS

Winning the environmental war is a whole lot tougher challenge by far than winning any other war in the history of man.

—Gaylord Nelson

Graph a Container

This activity will help you see how much solid waste you and your family add to the environment each week.

1. Choose one dinner you and your family had this week and write down all of the foods you ate. Beside each food, list all of the materials used to package that food.

EXAMPLE

Food	***Packaging Materials***
chicken noodle soup	*metal can*
	paper label
hamburgers	*polystyrene foam tray*
	clear plastic wrap
	paper label
french fries	*plastic bag*
ketchup	*glass bottle*
	metal cap
	paper label
frozen peas	*cardboard box*
	paper wrapper
salad	*plastic produce bags*
french dressing	*plastic bottle*
	plastic cap
	paper label
milk	*waxed cardboard carton*
brownies made from mix	*cardboard box*
	paper wrapper
	plastic bag
	cardboard egg carton
	plastic oil bottle

2. Make a bar graph showing the number of times each packaging material was used.

Packaging	**Number of Times**											
Material	**0**	**1**	**2**	**3**	**4**	**5**	**6**	**7**	**8**	**9**	**10**	**11**
Cardboard												
Glass												
Metal												
Paper												
Plastic												
Polystyrene foam												

3. Evaluate your results. Which materials were used most often?
4. Are the materials that were used most often biodegradable? Are they recyclable?
5. Similarly keep track of your dinners for an entire week and again graph your results.
6. You now have packaging figures for one week. What happens when you multiply these figures by the number of weeks in a year?
7. Your figures are for only one meal each day. What happens to these figures when you also consider the materials used to package the foods you eat for breakfast and lunch?

Smart Shopper's Checklist

How many of these things do you and your family do?

	WE BUY	INSTEAD OF
☐	cloth towels	paper towels
☐	wax paper	plastic wrap
☐	baking soda	air fresheners
☐	reusable metal knives, forks, and spoons	disposable plastic knives, forks, and spoons
☐	reusable ceramic plates, cups, and bowls	disposable paper plates, cups, and bowls
☐	food in glass jars with metal lids	food in plastic bottles or jars
☐	eggs in cardboard cartons	eggs in polystyrene foam cartons
☐	bottled drinks in glass containers	bottled drinks in plastic containers
☐	canned drinks packed in cardboard carriers	canned drinks packed in plastic rings
☐	toothpaste in tubes	toothpaste in pump dispensers
☐	bars of hand and bath soap	liquid soap in plastic pump containers
☐	cotton diapers	disposable diapers
☐	compact fluorescent light bulbs	incandescent light bulbs
☐	rechargeable batteries	nonrechargeable batteries
☐	phosphate-free, biodegradable dishwashing and laundry detergents	phosphate-containing, nonbiodegradable dishwashing and laundry detergents

Search for Recycled Paper

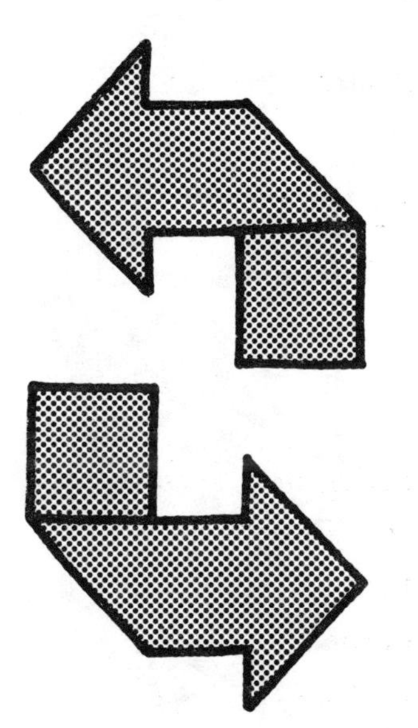

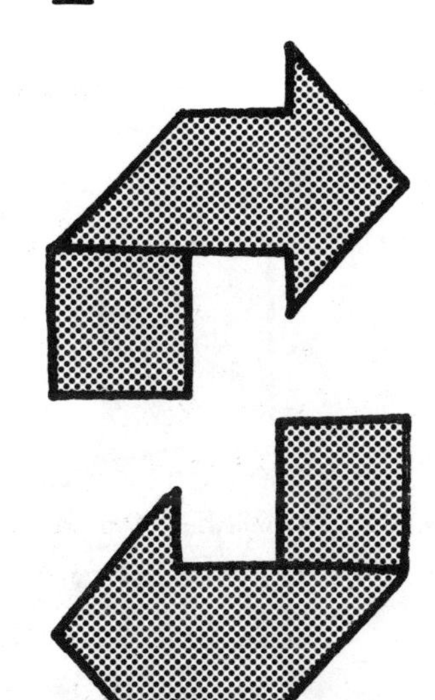

Recently, many companies have become more aware of the need to conserve our forests. These companies are making products from recycled paper and packaging products in recycled paperboard. Among the products currently being packaged in recycled paperboard are Cheerios cereal, Lipton Cup-A-Soup, and S.O.S scouring pads.

Search for packages and other products made from recycled paper in supermarkets, stationery and office supply stores, and greeting card shops. On a separate sheet of paper, list the ones you find. As clues in your search, look for the symbols and messages shown below.

PACKAGED IN RECYCLED PAPERBOARD

PACKAGED IN RECYCLED PAPERBOARD

This card is printed on 100% recycled paper.

corate a Box for Paper

Decorate a cereal box. Use it to hold sheets of paper that have been used on one side but are still clean on the other side.

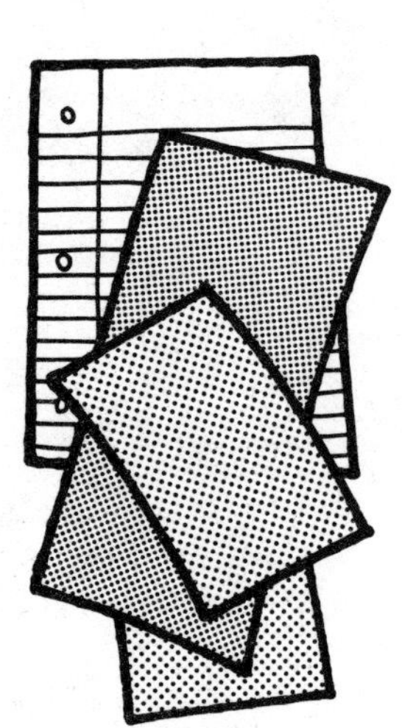

WHAT YOU NEED

- ☐ a large cereal box
- ☐ recycled gift wrap or brown paper cut from a grocery sack
- ☐ felt-tipped marking pens
- ☐ scissors
- ☐ tape
- ☐ glue

WHAT YOU DO

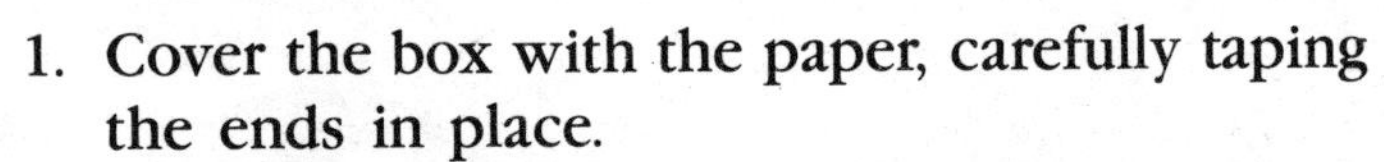

1. Cover the box with the paper, carefully taping the ends in place.
2. Label the box by cutting letters from paper scraps and gluing them to the box or by writing letters on the box with felt-tipped marking pens.
3. Decorate the box by drawing on it and/or by gluing paper shapes to it.
4. Use the box to hold art paper, computer paper, and/or writing paper that has been used only on one side.
5. When you need paper for a first draft, rough sketch, or shopping list, take a sheet from this recycled box and use the clean side.

VARIATION: Similarly decorate two larger cardboard boxes. Use one to store newspapers and the other to store nonglossy assorted paper that you cannot reuse but can recycle.

Wrapping Paper

The next time you are invited to a birthday party, instead of buying gift wrap, decorate your present in one of these ways.

Wrap it in a page from the comic section of the Sunday paper. This section is colorful, and the comic strips are fun to read!

Wrap it in a recycled brown paper grocery bag. Cut open the bag along one fold. Cut all the way around the bottom panel and remove it so that the paper will lie flat. Tape the paper around the box so that the *unprinted* side is out. Use crayons or felt-tipped marking pens to add designs and messages.

Wrap it in recycled wrapping paper.

Add a bow or ribbon you have saved.

Wrap it in an old wallpaper sample.

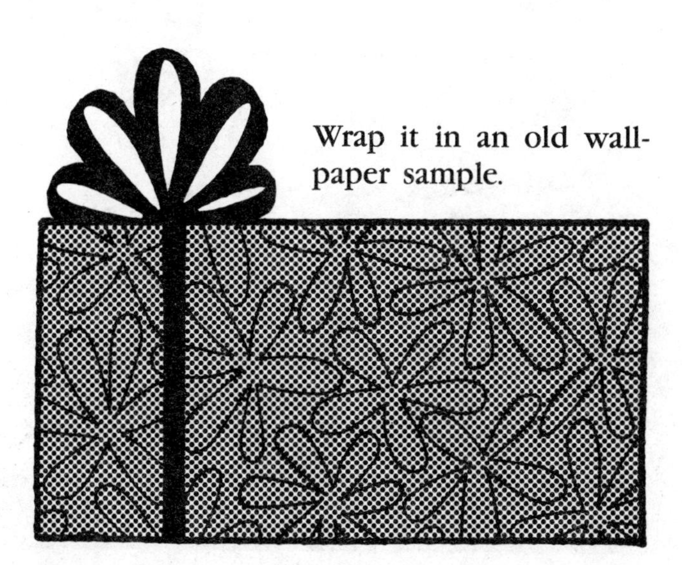

The Lunch Bunch

When you bring a lunch or snack to school, carry it in a lunch box or other reusable container. If you prefer paper bags, use the same one several times before throwing it away. Encourage your classmates to do likewise. For instructions and ideas about how to make a reusable fabric lunch bag, see pages 38–40.

How many students in your class bring lunches and/or snacks to school each day?

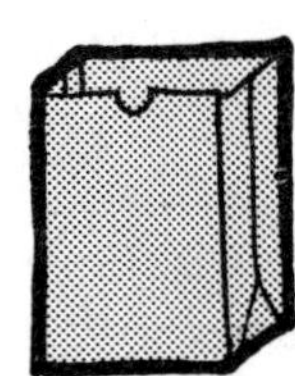

How many days are there in a school year?

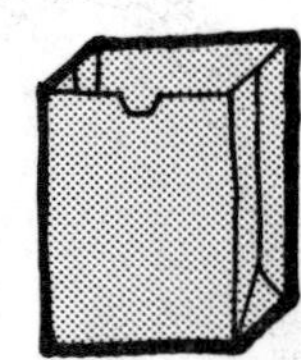

How many of these students usually carry their lunches or snacks in paper bags?

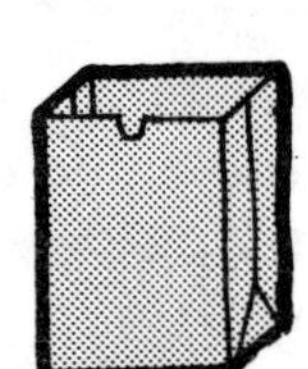

If all of the students who use paper bags bring new ones each day, approximately how many paper bags will the members of your class use in a school year?

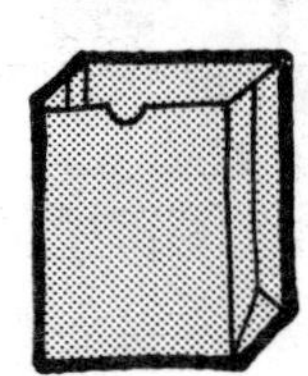

EARTHWORDS

The woods are made for the hunters of dreams,
The brooks for the fishers of song;
To the hunters who hunt for the gunless game
The streams and the woods belong.
—Sam Walter Foss
The Bloodless Sportsman

A culture is no better than its woods.
—W. H. Auden

Brainstorm a Bag

In how many ways can you recycle a used paper lunch bag? On a separate sheet of paper, list all of the ways you can think of to reuse one of these bags. Then pick a few of your favorite ideas and illustrate them.

VARIATION. Similarly brainstorm unusual or creative ways to reuse an aluminum can, a brown paper grocery sack, a cardboard box, or a plastic bag.

Paper Bag Rod Puppets—Head

Turn a paper lunch or grocery bag into a paper bag rod puppet. Then write a short skit about litter or recycling in which your puppet is one of the main characters.

WHAT YOU NEED

- ☐ a used paper lunch or grocery bag
- ☐ some newspapers
- ☐ a long cardboard tube
- ☐ a pair of scissors
- ☐ glue
- ☐ masking tape
- ☐ crayons or felt-tipped marking pens
- ☐ odds and ends, such as feathers, paper and fabric scraps, paper plates and cups (rinsed and dried for reusing), sticks, and yarn

WHAT YOU DO

1. Stuff the bag with newspapers.
2. Poke the cardboard tube into the bag to form the neck.
3. Squeeze the bag around the tube and use masking tape to fasten the bag securely around the tube.
4. With crayons or marking pens, draw a face on the bag.
5. Use some of your odds and ends to add a beak, a nose, ears, horns, and/or antlers.

YOU CAN MAKE

AN ELEPHANT with paper ears and a rolled-up newspaper trunk

A BEAR with a paper plate nose

A DRAGON with egg carton teeth

A BIRD with a cardboard beak and real feathers

A MOOSE with a paper cup nose and stick antlers

Paper Bag Rod Puppets—Body

After you have finished the head, you are ready to make the body of your paper bag rod puppet.

WHAT YOU NEED

- ☐ a piece of cloth about 2 feet square
- ☐ a pair of scissors
- ☐ a piece of cardboard
- ☐ glue
- ☐ stapler and staples
- ☐ a thin dowel about 15 inches long
- ☐ masking tape

WHAT YOU DO

1. Fold the cloth in half.
2. With the scissors, cut a slit in the fold for the tube as shown.
3. Make cardboard hands for your puppet.
4. Staple or glue the hands inside the folded cloth.
5. Put the tube, or neck portion, of the puppet's head through the slit in the cloth.
6. Glue the edge of the cloth around the puppet's neck to hold it in place.
7. Tape the dowel to one of your puppet's hands.

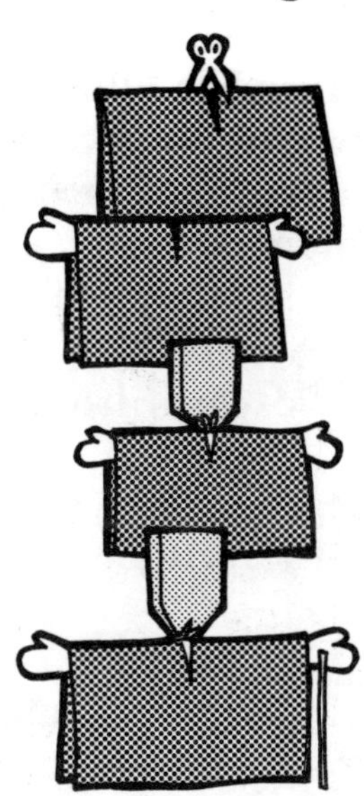

To move your puppet, hold the tube in one hand and the dowel in the other.

Fabric Lunch Bag Fun

Re... of those paper lunch bags you counted ... page 34? One way to save trees and energy and to reduce the amount of trash being produced each day is to carry your lunch or snack to school in a reusable container. On pages 38–40 are step-by-step instructions for a personalized fabric lunch bag you can have fun making and decorating. Because these directions are fairly lengthy, you might want an adult to sew the bag and leave the decorating to you.

WHAT YOU NEED

- ☐ ⅜ yard of corduroy, denim, or any other sturdy, washable, wrinkle-resistant fabric (Depending on the width of the fabric, this may be enough to make more than one lunch bag, so you can share with a brother, sister, or friend, or have an extra bag to carry when your first one needs washing.)
- ☐ sewing scissors and straight pins
- ☐ thread in colors to match and contrast with your fabric
- ☐ a needle if you plan to sew by hand or a sewing machine with an adult standing by to help
- ☐ some ways of decorating your bag (Use embroidery thread, fabric appliqué, scraps of felt and rick-rack, badges you have earned, patches you have purchased, fabric glue, and/or fabric paint.)
- ☐ an iron
- ☐ one strip of Velcro ¾ inch wide and 4½ inches long for each bag
- ☐ a pencil or piece of chalk with which you can mark on your fabric
- ☐ a piece of cardboard (optional)

WHAT YOU DO

1. From your large piece of fabric cut a smaller piece measuring 13 inches by 34 inches.
2. Decorate your fabric. Embroider letters or pictures on it, glue or sew fabric to it, or paint on it. To ensure that your decoration will be visible after the bag has been sewn, fold the fabric so that you have two 13-inch-by-17-inch halves. Separately decorate each half, leaving generous borders all the way around your decoration as shown. After decorating, let the paint or glue dry and then press as needed.

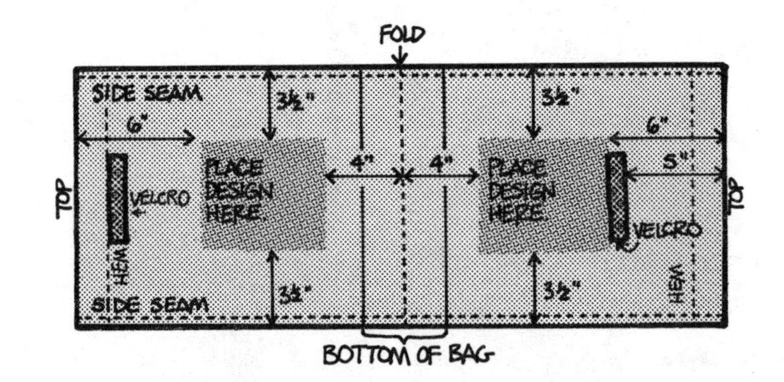

Fabric Lunch Bag Fun
(continued)

3. Finish the edges by sewing all the way around in a zig-zag stitch or using Fray Check.

4. Lay the fabric right side up and place the soft, or loop, portion of the Velcro strip so that it is 5 inches down from one of the 13-inch edges and runs parallel to that edge.

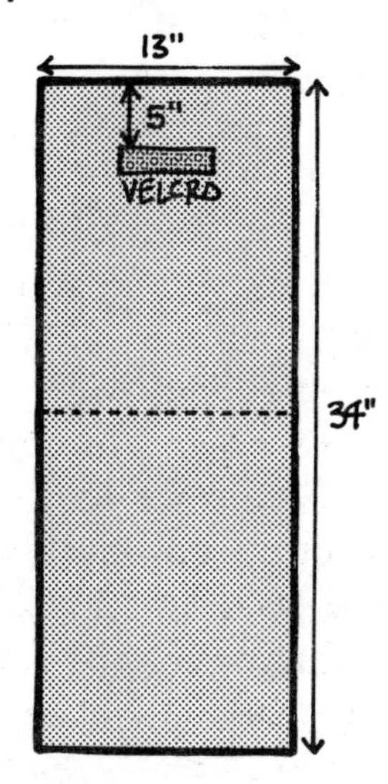

5. Center the Velcro strip, pin it in place, and sew all the way around it, keeping your stitches approximately ⅛ inch from the outer edge.

6. Fold the fabric in half (right sides together; wrong side out) so that the 13-inch edges meet at the top.

7. Pin the sides together.

8. Sew the sides together, stitching from the top (open end) to the bottom (fold), ½ inch in from the edge.

9. Press open the seams.

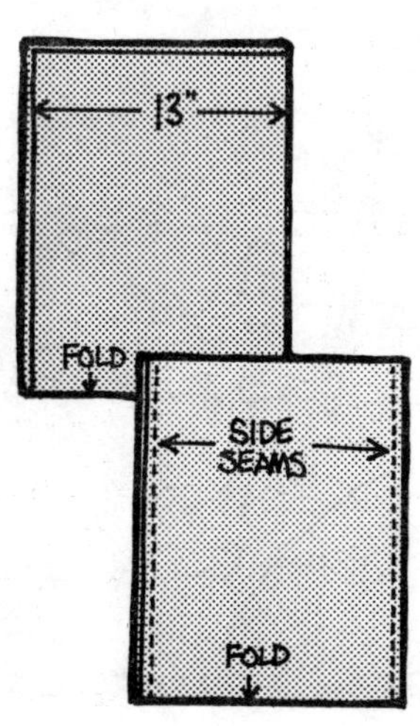

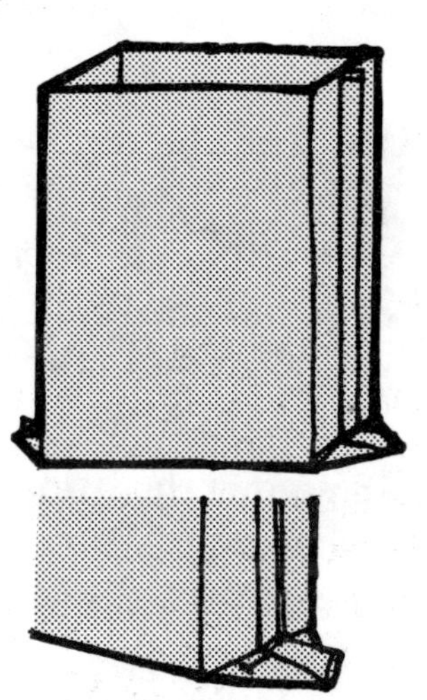

10. Shape the bottom of the bag by flattening the fold and pulling the two corners out to form triangles as shown.

11. Pinch these fabric triangles together and pin them, making certain that the side seam is in the center of each one.

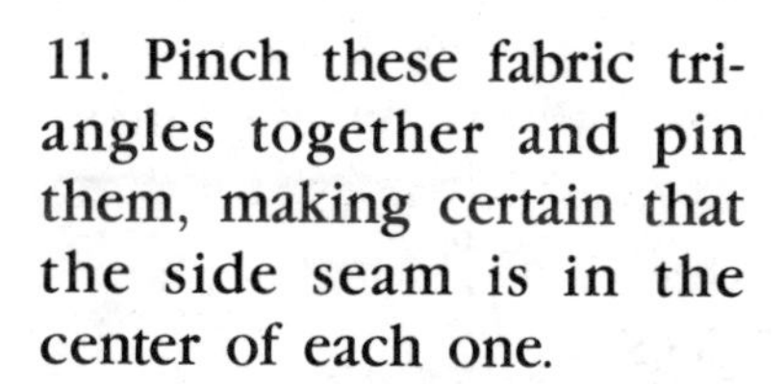

12. Using the pencil or chalk, carefully draw one line across the bottom of each triangle, 2 inches from the point.

13. Sew on this line and press.

14. With the bag still wrong side out, finish the top. Fold over and press ½ inch of fabric. Fold over an additional one inch of fabric and press again.

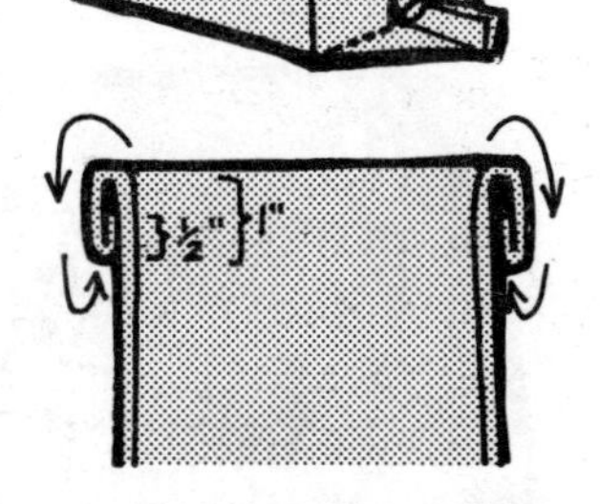

15. Sew around the folded top, ¾ inch from the edge.

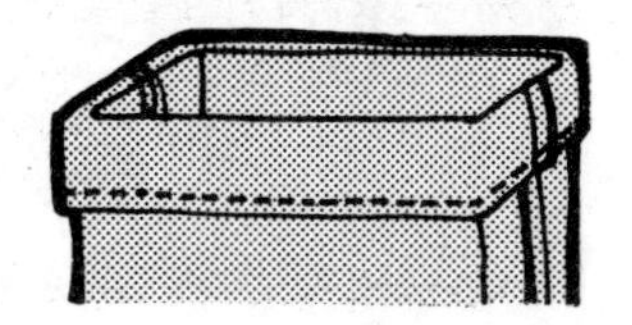

Fabric Lunch Bag Fun
(continued)

16. Turn the bag right side out.

17. Place the hard, or hook, portion of the Velcro strip at the top edge of the bag, on the side that does *not* already have a Velcro strip. The Velcro will align and fasten when the top of the bag is folded over twice.

18. Center the strip, pin it in place, and sew as you did in step 5.

19. To give your bag a rectangular shape, form edges by folding vertically from one of the bottom corners to the top of the bag, parallel to the side seam and 2 inches away from it.

20. Pin this fold in place and press.

21. Repeat steps 19 and 20 to form three additional edges.

22. Fold, pin, and press along the two longer edges of the bottom of the bag.

23. Topstitch along each of these six folds, ⅛ inch from the edge. You may want to use thread in a contrasting color and make these stitches part of your decoration.

24. (*optional*) Measure the bottom of the bag. Cut a piece of cardboard slightly smaller than these measurements and place it inside the bag to give it more shape and hold the bottom flat.

Toilet Paper Survey

You can help the environment when you buy toilet paper by doing one or more of the things listed below.

- **Switch from colored to white toilet paper.** Some of the dyes used to color paper produce **toxins**, substances that are poisonous to the environment, when these papers decompose.
- **Buy unscented toilet paper.** Some of the perfumes used to scent toilet paper also produce toxins.
- **Buy toilet paper that is packaged in paper** rather than in plastic.

- **Buy toilet paper made from recycled paper.** This type of toilet paper can usually be found in wholesale houses and supply stores that sell paper goods in bulk to hotels, motels, and restaurants.

Conduct your own toilet paper survey. Make a chart like the one shown below. Take it with you when you go to the supermarket. Examine five brands of toilet paper. For each one, place check marks in the appropriate boxes and record the price per roll.

Based on the results of your survey, which brand is the best buy for the consumer? Which brand is the best buy for the environment? Are these two "best buys" the same? Why or why not?

TOILET PAPER SURVEY CHART

Brand Name	Paper Wrapper	Uses No Dyes	Uses No Perfumes	Is Made From Recycled Paper	Price Per Roll

Be a Newspaper Saver

Encourage family members and neighbors to save old newspapers instead of throwing them away. Tie the newspapers with string or pack them in paper grocery bags. When you have gathered enough newspapers to make the trip worthwhile, take them to a recycling center.

Recycling newspapers

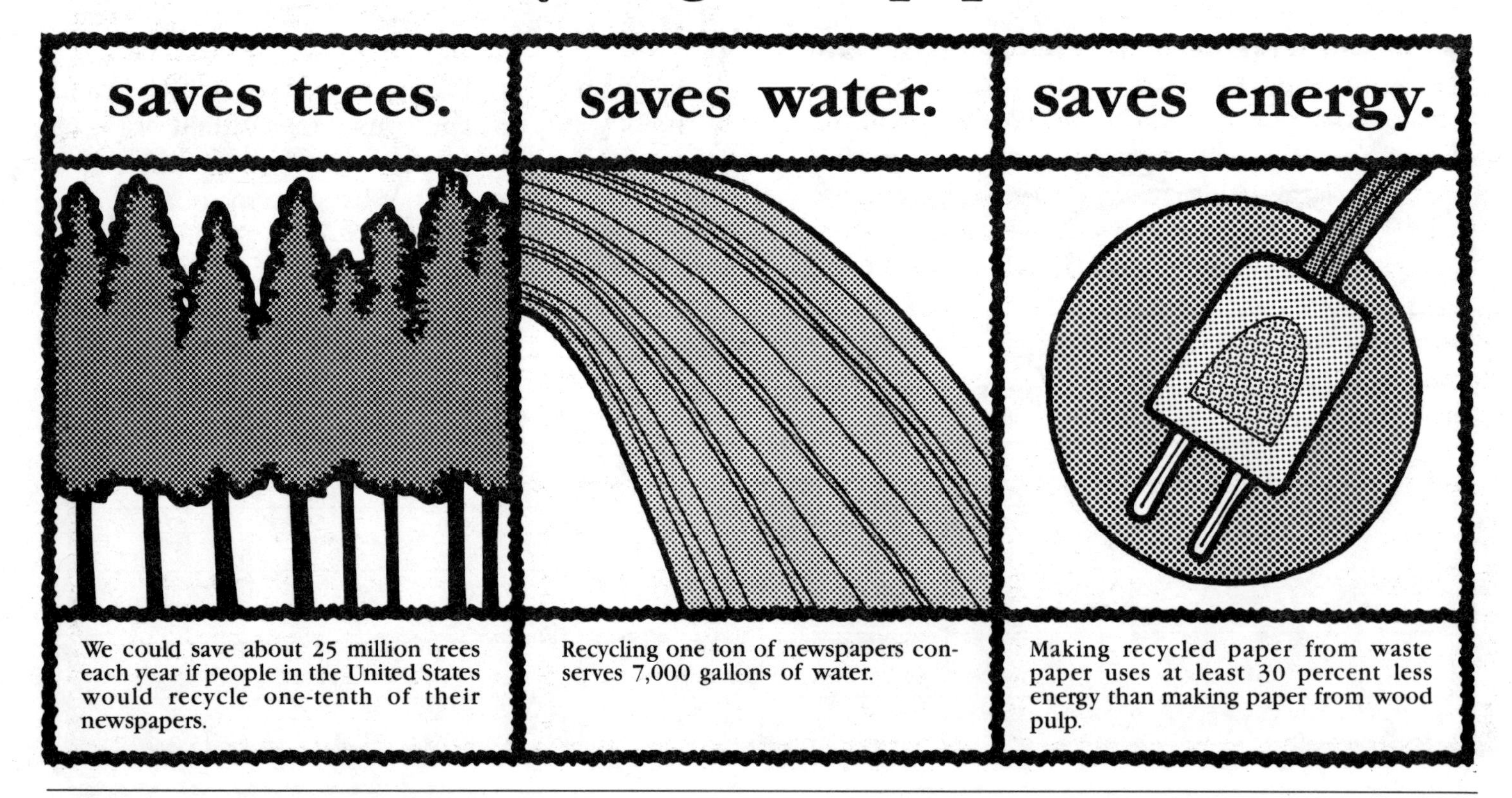

About Plastic

Plastic is the name given to any of a large group of substances made chemically from such materials as coal or oil mixed with water and limestone. Nylon, vinyl, and many cellulose products are plastics.

Plastic can be hard or soft, solid or clear. It can be molded by means of heat and pressure into a wide variety of forms, including fibers, sheets, and containers. Because of its versatility, plastic is often used instead of glass, metal, or paper.

As a packaging material, plastic offers some definite advantages. Plastic preserves freshness better than paper. It is sturdy and resists breaking better than glass. Also, containers made of plastic weigh less than same-size containers made of glass so they are easier to carry and cost less to ship.

But there are some disadvantages associated with plastic. It is made from coal or oil, which are **nonrenewable resources**. Plastic is very difficult to recycle. Currently, only about one percent of all discarded plastic is recycled. Also, plastic takes hundreds of years to decompose. The plastic we throw away today will linger in landfills for centuries.

You probably have many objects made from plastic around your home. Can you find at least twenty-five? Draw or describe them on a separate sheet of paper.

Fact

- **Americans throw out 2 billion disposable plastic razors each year.**

Fact

- **Before plastics can be recycled, they must be separated according to the type of resin used to make them.**

Fact

- **Recycled plastic is used to make toys, carpet backing, and fiber filling for ski jackets and sleeping bags.**

Piles of Polystyrene

Polystyrene is the main ingredient in foam plastics. Foam plastics can be easily molded into a variety of shapes. They are used to make picnic plates and cups, egg cartons, food storage boxes, meat trays, and packing materials.

Polystyrene foam seems like an ideal material for these uses. It is cheap, lightweight, and durable; but it does have some disadvantages. Polystyrene foam does not decompose. Once thrown away, it lingers in landfills forever. Also, as polystyrene is being made, chemicals called **chlorofluorocarbons** (CFCs for short) are released into the air.

Scientists believe that CFCs thin the **ozone layer**. This layer is formed naturally, high above the earth, in the upper atmosphere. It protects plants and animals from the sun's harmful **ultraviolet rays**. A thinner ozone layer means less protection.

During the 1980s, people became more and more concerned about the effects of CFCs on the ozone layer. In response to this concern, polystyrene manufacturers voluntarily agreed to quit using CFCs in the production of food-service products.

Fact

- **Polystyrene products rank fifth among all U.S. products in the amount of toxic waste created while they are being made.**

Fact

- **Polystyrene foam from a single day's sale of hamburgers by just one fast-food chain takes up to 50,000 cubic feet of landfill space.**

Fact

- **When polystyrene products burn, as many as 57 chemicals are released into the air.**

Paper or Plastic?

At most supermarkets, you can choose whether to have your groceries sacked in paper bags or in plastic bags. Do some research to discover the advantages and disadvantages of each type of bag. Find out which one is made from a renewable resource, which one can be recycled, and which one is biodegradable. In doing your research, consider not only the effects of these bags on the environment but also the convenience, safety, and cost associated with their use. Summarize your findings on a chart similar to the one shown below.

For health reasons, supermarkets cannot permit one customer to return paper or plastic bags for use in sacking groceries bought by other customers; but many supermarkets will allow you to bring in and reuse your own bags. Or, if you prefer, you can purchase reusable cloth shopping bags in health food stores or order them through catalog shopping services.

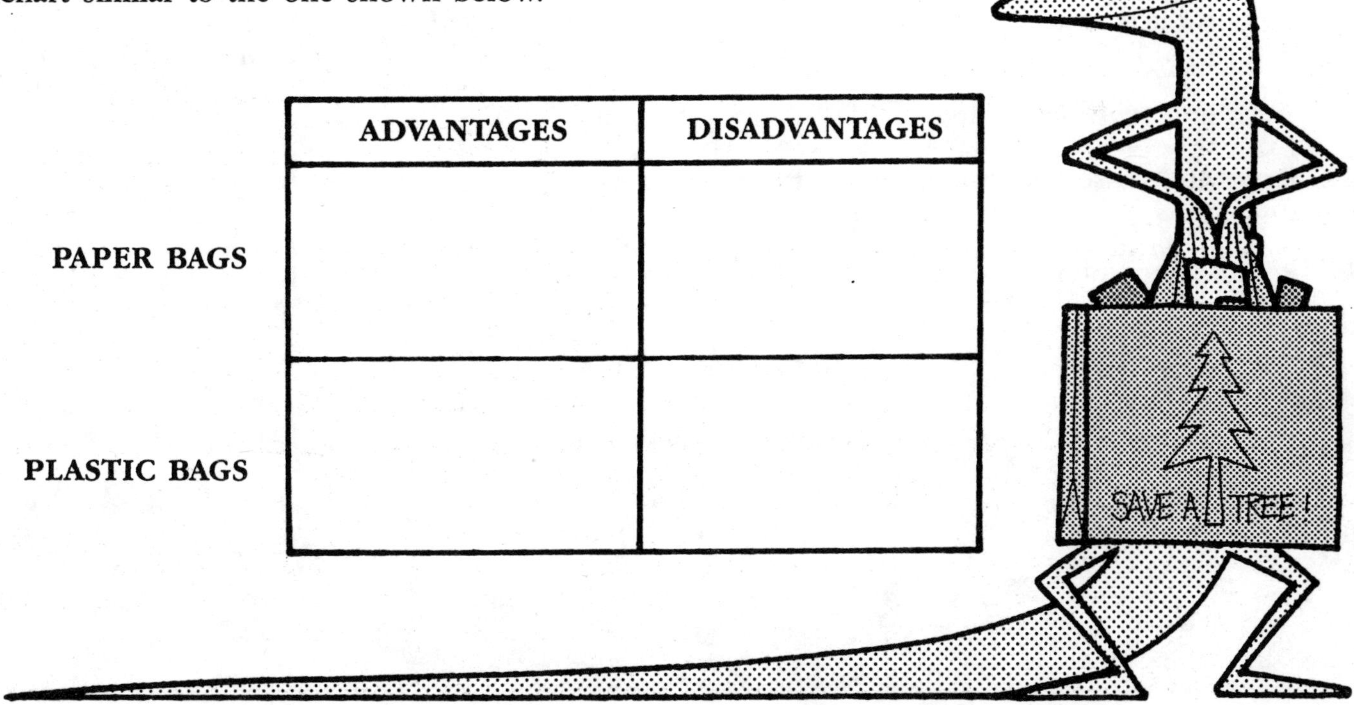

	ADVANTAGES	DISADVANTAGES
PAPER BAGS		
PLASTIC BAGS		

Sixteen Simple Steps

Here is a list of sixteen simple steps you can take to help heal the environment. Put a check mark in the box beside each step you take during a two-week period. Make these steps part of your routine.

PACKAGING

- ☐ Select products that come in biodegradable packages whenever possible.
- ☐ Buy products in returnable bottles, and return them.
- ☐ Wash and reuse glass jars.
- ☐ Rinse and reuse aluminum foil.
- ☐ Recycle aluminum cans.

PAPER PRODUCTS

- ☐ Use cloth napkins instead of paper ones.
- ☐ Buy greeting cards that have been printed on recycled paper.
- ☐ Reuse gift wrap.
- ☐ Recycle newspapers.
- ☐ Write on the *back* of a sheet of notebook paper, not just on the front.
- ☐ Reuse paper lunch bags or carry your meal in a fabric bag or lunch box.
- ☐ Save and reuse cardboard gift and shipping boxes.

PLASTIC PRODUCTS

- ☐ Select nonplastic products whenever possible.
- ☐ When you shop in grocery stores or supermarkets, avoid putting fruits and vegetables in plastic bags.
- ☐ Rinse out and reuse plastic produce and grocery bags.
- ☐ Encourage your local recycling center to begin accepting plastic.

A Week's Worth of Waste

Many of the items you buy are packaged in plastic. You probably open each package, use what is inside, and throw the plastic away without giving it much thought. But what if you saved all of this plastic? How much plastic waste would you accumulate in one short week?

TRY THIS

1. Obtain an empty cardboard box.

2. Weigh this empty box on a bathroom, kitchen, or postal scale. Write down the weight.

3. Collect *all* of the plastic waste produced in your home for a week. Include knives, forks, spoons, cups, and containers, as well as bags, boxes, and wrappings.

4. Wash these plastics and set them out to dry.

5. Crush your clean plastics. If some of them are six-pack rings, carefully cut open all of the loops.

6. Put your crushed plastics in the cardboard box.

7. At the end of the week, weigh this box of plastics on a bathroom, kitchen, or postal scale. Write down the weight.

8. Subtract the weight of the empty box from the weight of the plastic-filled box to determine the weight of the plastic. How much plastic waste did your family accumulate in a week?

9. Use the amount of plastic waste your family accumulated in a week to estimate how much plastic waste your family throws out in a month and in a year.

Aluminum Can Drive

Many of the aluminum cans you find in the grocery store are made of recycled materials. Fifty-five percent of all the aluminum cans produced in 1988 were recycled by consumers. With some of your classmates, organize a school-wide drive to collect and recycle aluminum cans.

1. Plan the drive. When and where will it be held? What adults and kids will help? Who will provide containers to hold the cans you collect? Who will take these cans to a recycling center? How will the money you receive be spent?
2. With the help of a teacher or principal, schedule drive dates and designate a place on the school grounds or in your school parking lot as the official collection site.
3. Contact a recycling center that handles aluminum to learn about its hours of operation. Tell center personnel about the aluminum can drive you are planning and ask them for tips about how best to prepare aluminum for recycling. The center may have large collection containers you can borrow and/or fliers you can distribute.
4. Make posters to encourage students to save their aluminum cans and bring them to school for recycling.
5. Design a flier students can take home to let family members and neighbors know about the drive.
6. Hold the drive and collect the cans. Remind volunteer workers to keep the collection site neat and to thank *everyone* who contributes.
7. Take the cans to the recycling center.
8. Use any money you receive in whatever way you decided before the drive. If you are short on ideas, consider donating it to an organizaton that is working to clean up the environment or giving it to your school with the understanding that it will be used to buy plants for the school yard, books for the library, equipment for the playground, or something else many students will enjoy.

Tin Can Stilts

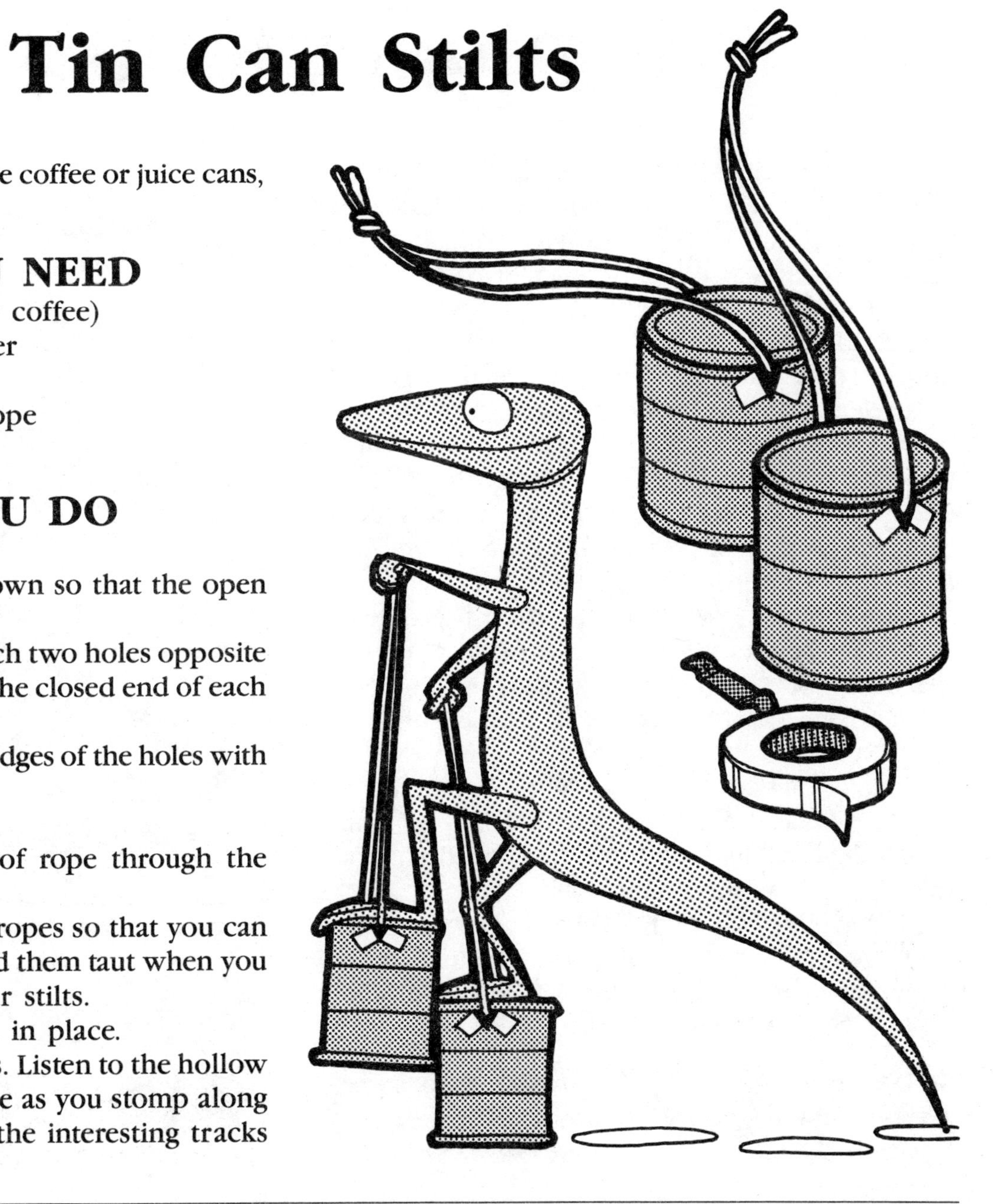

Instead of throwing away large coffee or juice cans, turn them into stilts.

WHAT YOU NEED

- ☐ two large cans (juice or coffee)
- ☐ a punch-type can opener
- ☐ masking tape
- ☐ 10 feet of lightweight rope
- ☐ scissors

WHAT YOU DO

1. Rinse and dry the cans.
2. Turn both cans upside down so that the open ends are at the bottom.
3. With the can opener, punch two holes opposite each other in the sides at the closed end of each can.
4. Carefully cover the sharp edges of the holes with masking tape.
5. Cut the rope in half.
6. Loop one 5-foot length of rope through the holes in each can.
7. Adjust the lengths of the ropes so that you can easily reach them and hold them taut when you stand up straight on your stilts.
8. Knot the rope to hold it in place.
9. Walk on your tin can stilts. Listen to the hollow clonking sound they make as you stomp along on the sidewalk. Notice the interesting tracks they leave in the dirt.

Cut Open Those Rings

The plastic rings used to package six-packs of soda are harmful to the environment in at least two ways. First, most of them are *not* biodegradable. As a result, they become permanent parts of our landfill problem.

Second, many of them end up floating in our ponds, lakes, streams, rivers, or oceans. Because these rings are transparent, diving birds do not see them. These creatures unwittingly plunge their beaks or heads through the rings. Unable to free themselves, the birds starve to death because they cannot open their beaks to eat, or suffocate because they cannot open their throats to breathe.

In an effort to deal with the landfill problem, many state legislatures have passed laws requiring that the rings on all six-packs sold within their borders be made of biodegradable plastic. This kind of plastic breaks down when it is exposed to ultraviolet rays from the sun. To determine if the plastic rings on soda six-packs sold in your state are biodegradable, look for a diamond design on the plastic near the finger hole.

Laws may be helping the landfills, but the water birds are relying on you. To protect them, carefully cut open *all* of the loops before you discard plastic six-pack rings.

EARTHWORDS

Man can no longer live for himself alone. We must realize that all life is valuable and that we are united to all life.

—Albert Schweitzer

Reuse Wire Coat Hangers

Don't throw away your extra wire coat hangers. Instead, sort them by type or shape, fasten them together, and return them to your dry cleaner for reuse. Or use your imagination and follow these simple rules to put surplus coat hangers to work in different and creative ways.

If you are stuck for ideas, try to create

- a back scratcher,
- a book rack,
- a card or letter holder,
- a fly swatter,
- a decorative mobile,
- a pen holder, or
- a plant hanger.

RULES

1. You may use more than one coat hanger.
2. You may bend or cut the coat hangers, but use pliers (*not* bare hands) and a good pair of wire cutters (*not* sewing scissors). This could be tricky, so ask an adult for help.
3. You may attach several coat hangers together.
4. You may attach other objects or materials to the coat hangers.

All About Glass

Glass is a very hard substance that breaks easily and can usually be seen through. It is made by melting sand with certain chemicals. While glass is very hot, it can be shaped into a variety of useful objects. Glass is used to make lenses and windows. It is also used to make containers, including bottles and jars.

Glass containers are used to hold many different liquids. Among them are foods, drinks, and cosmetics. Because glass is transparent, shoppers can see what is inside. Products with pleasing colors and textures retain their natural eye appeal when packaged in clear glass.

The glass containers used to hold food products are of two types, refundable and nonrefundable. **Refundable glass containers** can be reused commercially. Bottles and jars of this type are made of thick, heavy glass. They can be washed, dried, and refilled as many as thirty times. Water and other beverages are often sold in refundable glass bottles.

When you buy products in refundable glass containers, you are charged a **deposit** on each container. If you return these containers to the place of purchase, the deposit is refunded to you. Thus, refundable containers have a **redemption value**. This value is the amount of money that will be paid when one of these containers is bought back, or **redeemed**.

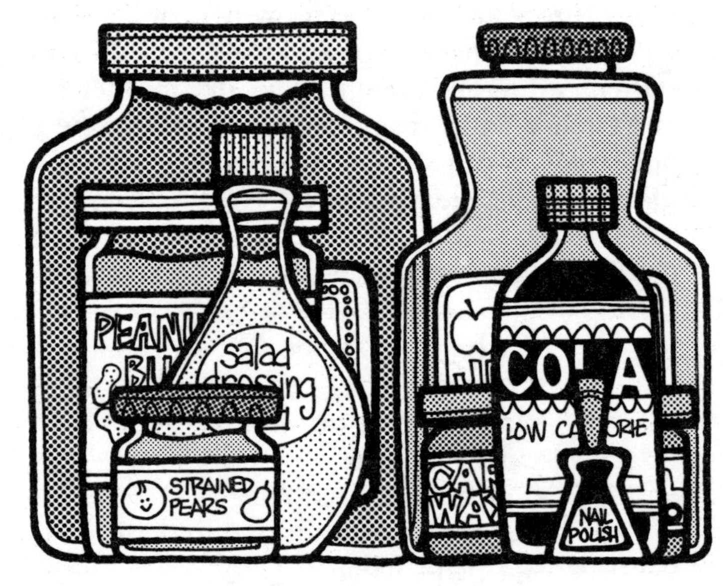

Fact

- **In the United States, more than 28 billion glass bottles and jars are thrown away each year.**

Fact

- **Every glass bottle or jar on the shelf in a supermarket contains about 20 percent recycled glass.**

Fact

- **Glass bottles make up 8 percent of the solid wastes that find their way into U.S. landfills.**

All About Glass
(continued)

Nonrefundable glass containers cannot be reused commercially. They are made of thinner glass than refundable glass containers. As a result, they are lighter weight and are much easier to carry; but they are not strong enough to withstand the thorough cleaning needed to make them safe for reuse.

When you buy products in nonrefundable glass containers, you are not charged a deposit. These containers have no redemption value, but you can reuse them in your home for mixing liquids and storing leftovers. Or you can take them to a glass recycling center where they will be melted down and made into new glass products.

Recycling glass is good for the environment because is keeps glass out of landfills and saves natural resources. But there are environmental costs associated with recycling. Glass will not melt unless it is heated to a temperature of about 2000° F. As you might imagine, making anything that hot takes a lot of energy!

TO RECYCLE A GLASS CONTAINER

1. Remove the cap, lid, or ring. (You need not worry about removing the paper label.)
2. Rinse out the container.
3. Dry the container or let it drain.
4. Place the container in a sturdy cardboard box.
5. When the box is full, take it to a glass recycling center.

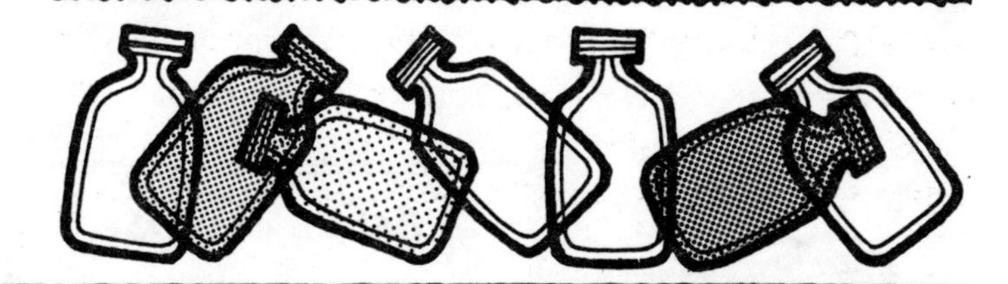

WHO?	WHAT?	HOW?
Who first made and used glass?	What natural gemstone is hard enough to scratch glass?	How is glass **blown** to make vases? How is it **ground** to make lenses?

Clean Up Your Room

Look through your closet, drawers, and shelves to find books, clothes, toys, and games you have outgrown, no longer want, and/or no longer use. Instead of throwing away all of these things and adding to the pollution problem, see how many of them can be reused or recycled. For example, you might give outgrown clothing to a younger child in your family or neighborhood. A church, synagogue, or child-care group might welcome some of your toys and games. Most school and public libraries will accept books if they are in good condition. Soft cotton knits that are too worn to wear make good dust cloths or paint rags. And don't forget to save and recycle the paper from old homework assignments and art projects, used coloring books, and the like. When you have finished, your room will look better and you'll have more space to store the stuff you use, want, and wear!

EARTHWORDS

Use it up, wear it out;
Make it do, or do without.
—New England maxim

A Clean-up Treasure Hunt

With a group of friends, conduct a clean-up treasure hunt on your school grounds. Not only will you have fun looking for the items listed, but you will pick up litter as well!

WHAT YOU NEED

- ☐ a group of friends to divide into teams
- ☐ two trash bags for each team
- ☐ work gloves to protect all hands involved
- ☐ five sturdy boxes labeled **glass**, **metal**, **paper**, **plastic**, and **organic**
- ☐ at least one copy of the Wanted list for each team

WHAT YOU DO

1. Divide the group into teams of two or more.
2. Give each team two trash bags, gloves, and a copy of the Wanted list.
3. Set a time limit.
4. Spread out and search for the items on the list.
5. As you find items, check them off the list, and carefully pick them up and place them in one of the bags.
6. Put any unlisted litter you find in the other bag.
7. When time is up, see which team has found the largest number of listed items and declare this team the winner.
8. Sort all of the litter into the labeled boxes.
9. Dispose of this litter properly. Return, reuse, or recycle what you can. Discard the rest by placing it in a garbage can or trash bin.

Wanted

a ballpoint pen	a pencil
a bottle cap	a blue object
a broken crayon	a red object
a candy wrapper	a yellow object
a lunch bag	something lost
a paper clip	something glass
a pencil	something metal
a piece of clothing	something plastic
notebook paper	something round
a rubber band	something square
a paper clip	a leaf or twig

EARTHWORDS

We won't have a society if we destroy the environment.

—Margaret Mead

What Is Energy?

Energy is the capacity to do work or the ability to make things move. It is an important part of your everyday life. Electrical energy coming into your home turns on lights, heats toasters, cools refrigerators, brings pictures to television screens, and runs the water pumps in dishwashers. Energy created by burning gasoline powers the engines of buses, cars, trucks, and vans.

Energy is available in two forms. One of these forms is called **kinetic energy**. It is the active energy found in heat, light, sound, and motion. The other form is termed **potential energy**. It is the stored energy available in resources such as coal, gasoline, oil, and water held behind a dam.

To make use of potential energy, we must change, or **convert**, it to kinetic energy. Thus, to create hydroelectric power, the water behind a dam is released. As this water falls, its energy becomes kinetic. It is able to push the blades of giant turbines and, by doing so, to produce electricity.

The process of converting energy from one form to another often results in by-products we really don't want and sometimes don't know what to do with. For example, when fuels are burned to produce energy, unwanted chemicals are released into the air. And when atoms are split to produce nuclear energy, radioactive waste is produced and must be disposed of.

Fact

- **The United States uses more energy per person than any other nation in the world.**

Fact

- **This dependence on large amounts of energy increases U.S. vulnerability to oil spills and other disasters.**

Fact

- **Energy production is one cause of air pollution and acid rain.**

Energy Facts in Brief

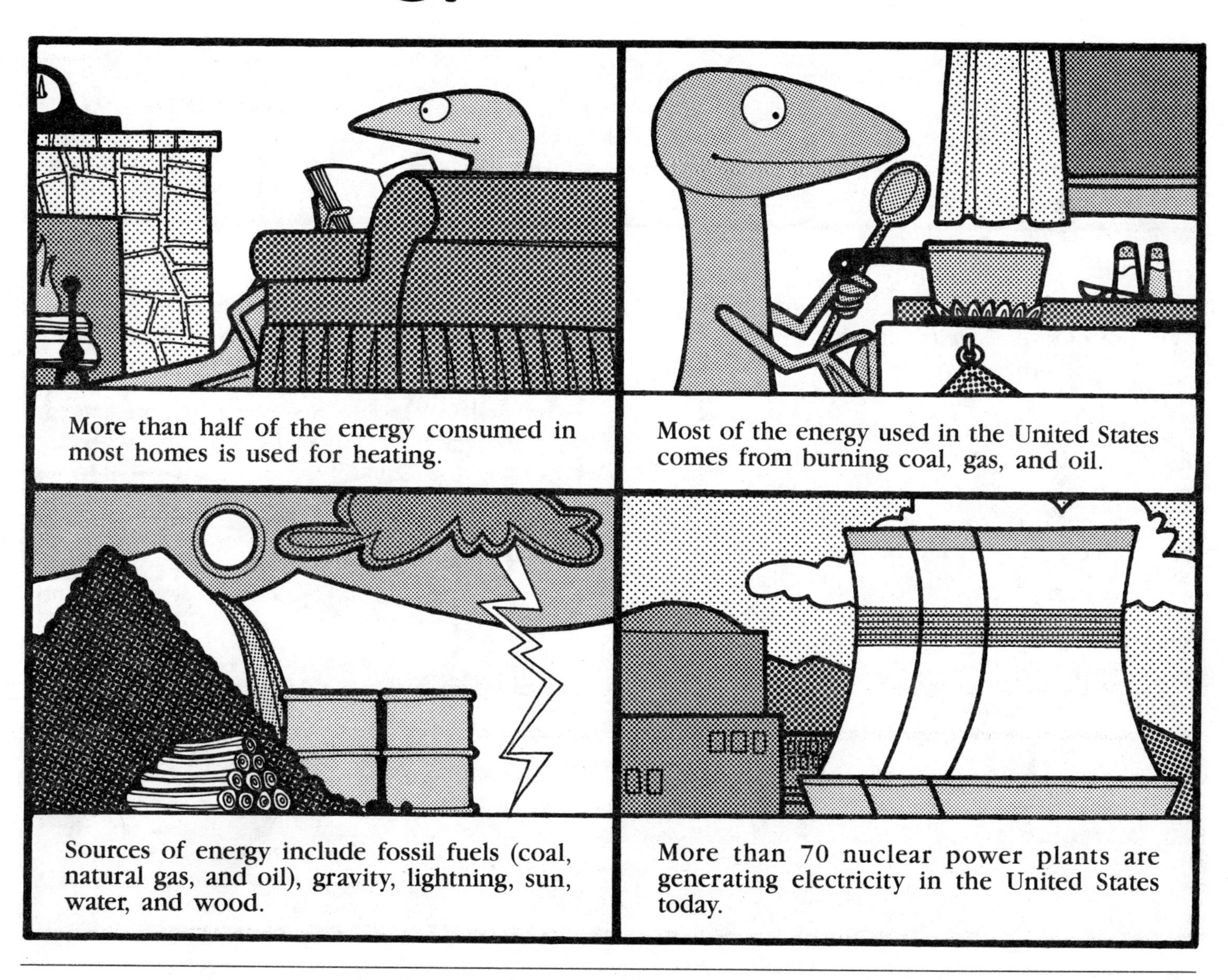

Checklist Changes

Fact	Fact	Fact
• 40 percent of the energy you use in your home is for heat.	• Of the electricity generated in the United States, 20 percent is used to keep the lights on.	• A microwave oven uses one-third to one-half as much electricity as a conventional oven.

Determine which ones of your major household appliances use electricity. (Some of your appliances may use only electricity, others may use natural gas, and still others may use a combination of electricity and gas.) Check your electrical meter, and record the reading and the date. (See page 60 for instructions on how to read a meter.) Talk with members of your family about changes you can make to cut back on your household use of electricity. See the Family Electricity Checklist on page 59 for ideas. Make some of these changes. After a month, read your electrical meter and compare your new electricity bill with older ones. Figure out approximately how many kilowatt-hours of electricity and how much money you saved as a result of the changes you made. Show these savings on a chart or graph.

Fact	Fact	Fact
• Every year Americans buy more than one billion incandescent light bulbs.	• A 60-watt incandescent light bulb will last about 750 hours.	• A 20-watt compact fluorescent bulb will generate the same amount of light as a 60-watt incandescent bulb and last about 7,500 hours.

Family Electricity Checklist

- ☐ Set the thermostat on your refrigerator at 38 ° F.
- ☐ Set the thermostat on your hot water heater no higher than 120 ° F.
- ☐ Set the thermostat on your furnace between 65 ° and 70 ° F. Instead of turning up the furnace when the house feels cool, put on a sweater.
- ☐ Set the thermostat on your air conditioner between 75 ° and 80 ° F. Use well-placed, energy-efficient fans to circulate the air and make the house feel cooler than it is.
- ☐ Avoid heating or cooling unused rooms in your house. Close the doors and ducts to rooms that are not in use.
- ☐ Replace high wattage bulbs with lower wattage ones where bright light is not needed.
- ☐ Turn off the lights when you are the last person to leave a room.
- ☐ Replace inefficient incandescent light bulbs with more energy-efficient compact fluorescent bulbs.
- ☐ Keep refrigerator doors closed. Don't leave them open while you empty or fill ice trays or while you think about what you want to eat or drink.
- ☐ As older household appliances wear out, replace them with energy-efficient ones.
- ☐ Run the dishwasher only when it is full.
- ☐ Operate the clothes washer only for full loads.
- ☐ Whenever possible, use cool water instead of warm or hot water.
- ☐ Clothes dryers operate more efficiently when the lint screen is clean. Remember to check the lint screen after each load and clean it whenever necessary.

Be a Meter Reader

Time is measured in hours. Electricity is measured in watts and kilowatts. (One thousand watts make one kilowatt.) Electrical energy is measured in kilowatt-hours. A **kilowatt-hour** is the amount of energy present in one kilowatt of electricity supplied for one hour of time. One kilowatt-hour is enough electrical energy to keep a standard 100-watt bulb burning for 10 hours.

A meter is attached to the line that brings electricity to your home. This meter measures the electrical energy you and your family use. An employee of the power company serving your area reads this meter regularly. These readings determine the amount of money your family is charged for the electrical energy it has used.

Charges for electrical energy are based on a set price per kilowatt-hour. Some power companies have established a **baseline allowance** for residential customers. This allowance is the amount of electrical energy needed to heat space and to operate essential appliances and equipment. These companies charge a lower rate for kilowatt-hours within the baseline allowance than for kilowatt-hours that exceed this allowance.

To find out how much electrical energy your family uses, be a meter reader. (See the tips in the box.) Write down the date and the numbers shown on the meter dials. Read the meter again one week later. Subtract the older reading from the newer one to determine how many kilowatt-hours of electrical energy your family used in a week.

To find out how much your family will be charged for the electrical energy it has used, call the power company that serves your area. Ask how much it charges for each kilowatt-hour. Then, multiply your weekly total by this amount. In general, power companies charge residential customers between 5 cents and 15 cents for each kilowatt-hour they use.

Tips for Meter Readers

- **Look at your electrical meter. Notice that some dials are numbered in a clockwise direction while others are numbered in a counterclockwise direction.**
- **Read the dials from *right to left*. The first dial shows kilowatt-hours. The second dial shows tens of kilowatt-hours. The third dial shows hundreds of kilowatt-hours, and so on.**
- **Record the number each needle points to, writing from *right to left*. If a needle points between numbers, write down the smaller number.**

All About Batteries

A **battery** is a container holding materials that produce electrical energy by chemical reaction. Many of the chemicals used in batteries are highly toxic. Batteries have limited lifetimes. Worn-out batteries are often carelessly discarded. They end up in landfills. There, they split open and leak their poisonous contents into the soil.

Batteries can be disposable or rechargeable. **Disposable batteries** stop making electrical energy when the chemicals in them can no longer react. These batteries may last only a few months and then be thrown away. **Rechargeable batteries** are designed to store energy for later use. They come with a recharger unit that can be plugged into any standard electrical outlet. When these batteries run out of power, they can be recharged and used again.

Although rechargeable batteries cost more, they are a better buy than disposable batteries because they last far longer and are much kinder to the environment.

Ways You Can Help

- **Turn off all battery-operated toys when you are not playing with them so that you will not waste the power in their batteries.**
- **When possible, buy things that don't need batteries. If you need a calculator, consider purchasing one that is solar-powered.**
- **When you do need batteries, buy rechargeable ones rather than disposable ones.**

Batteries are often used to provide power for toys, calculators, cameras, clocks, flashlights, smoke alarms, and portable radios and tape players. Check your home or apartment for battery-powered devices, and make a list of the ones you find.

Fact

- **More than 2 billion batteries are sold in the United States each year.**

Fact

- **Among the chemicals used in batteries are lead, mercury, sulfuric acid, and zinc.**

Fact

- **Batteries can be as small as a dime or as big as a shoe box and weigh as little as a fraction of an ounce or as much as 30 pounds.**

Wind Energy

We rely on fossil fuels such as coal and petroleum to provide much of the energy needed to heat homes, power vehicles, and move machines. It took nature millions of years to create this fuel supply. Now it is dwindling rapidly and may soon run out.

One alternate source of energy is the wind. In several naturally windy places in California, scientists are experimenting with **wind farms**. These farms are acres of relatively flat or gently rolling land on which people build rows of windmills instead of planting rows of crops.

A **windmill** is a mechanical device that consists of blades, or rotors, attached to a central pole. The energy created when the wind turns these blades is being used to generate electricity. This electricity is sold to utility companies and to consumers.

Burning fossil fuels to create heat and energy reduces our supply of these fuels and results in byproducts that pollute the air. Using wind in these ways does not diminish the supply of wind or dirty the air.

SOMETHING TO THINK ABOUT

Wind farms might seem to be one ideal solution to the energy problem. But, as with other solutions to this problem, the use of wind farms is not without drawbacks. Land used in this way may not be available for farming, housing, or recreation. Find out how many acres of land must be devoted to wind farms to generate a substantial amount of energy. Then evaluate the use of the land in this way. Is the gain in energy worth the loss in land? Why or why not?

Solar Energy

Solar energy is energy produced by the sun. Sunlight can produce enough heat to start a fire and enough electrical power to operate all of the instruments aboard a spacecraft. To understand how sunlight produces heat, think about how hot the air inside a car becomes when it is parked in the sun with the windows rolled up on a bright, clear day. Sunlight can be changed into electrical power by devices called **solar cells**. This power is used to operate a variety of electrical appliances, including calculators, fans, and flashlights.

Solar energy offers some advantages over other forms of energy. It is clean. Making it does not pollute the air. Using it does not result in smog, smoke, soot, or any of the chemicals that cause acid rain. Also, solar energy is produced from sunlight, which is a renewable resource, not from fossil fuels, which are nonrenewable resources.

Although solar energy is clean and is produced from a renewable resource, there is at least one problem associated with its use. Solar energy is not always readily available when it is most needed. As you might imagine, the amount of solar energy that can be produced depends on the amount of sunlight. For this reason, less solar energy is produced on cloudy days than on clear ones, and no solar energy is produced at night. But more energy for heat is needed on cloudy days than on clear ones, and more energy for light is needed at night than during daylight hours. Thus, to make the sun a convenient and dependable energy source, ways must be found to store solar energy when it can be most easily produced for use when it is most needed.

Using your creative imagination, invent a device that stores or is powered by solar energy. Draw and label a diagram of your device. Write a paragraph or two telling how it works. List the special features that make your device good for the environment.

Fact

The tremendous light and heat of the sun are produced by burning 4½ million tons of hydrogen gas each second.

Fact

Groups of solar cells arranged together on **solar panels** have provided electrical power for the instruments on manned and unmanned spacecraft.

Fact

Solar panels provide the power needed to operate emergency telephones along some California freeways.

Bright Ideas to Write About

Environmental Coat of Arms

Create an environmental coat of arms.

1. Select one of these traditional shapes and draw it on a sheet of paper.

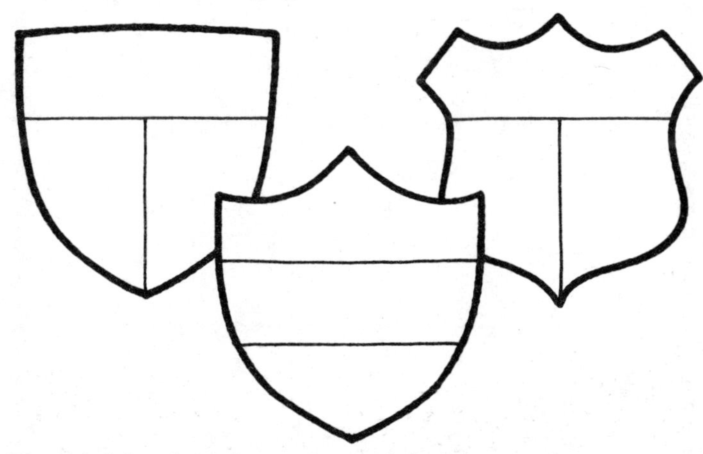

2. Divide the shape into three sections.
3. In one section, draw a picture or use a symbol to represent something in the environment you care about.
4. In another section, draw a picture or use a symbol to represent something that poses a threat to the thing you care about.
5. In the third section, draw a picture or use a symbol to represent something that can be done to protect the thing you care about from this threat.
6. In the banner below your coat of arms, write a motto, a few words that tell what the pictures or symbols on your coat of arms represent.

EARTHWORDS

In wildness is the preservation of the world.

—Henry David Thoreau

An Energy Contract

First, think of three things you can do regularly to conserve energy during a two-week period. Second, write an energy contract in which you promise to save energy by doing these three things. In writing the contract, follow the example shown on this page or create your own. Next, ask your mom or dad to sign your contract. Then, do what you promised.

At the end of two weeks, evaluate your results. Were you successful in meeting the energy-saving goals you set for yourself? What difficulties did you encounter? Finally, see if you can make these three energy-saving ideas and others like them a regular part of your own earth-friendly life-style.

Ideas

- Turn off the lights when you are the last person to walk out of a room.
- Walk or ride a bicycle to school instead of being driven in a car.
- If you cannot walk or ride a bicycle to school, join a carpool or take a bus.
- Use a manual pencil sharpener instead of an electric one.
- Use a manual can opener instead of an electric one.

Energy Contract

I, ________________, do promise that I shall help to conserve energy for a period of two weeks by

1. ______________________________
____

2. ______________________________
____

3. ______________________________
____

____
(signature)

____
(date)

____
(parent's signature)

AIR,
LAND,
& WATER
H
POISON

Facts in Brief

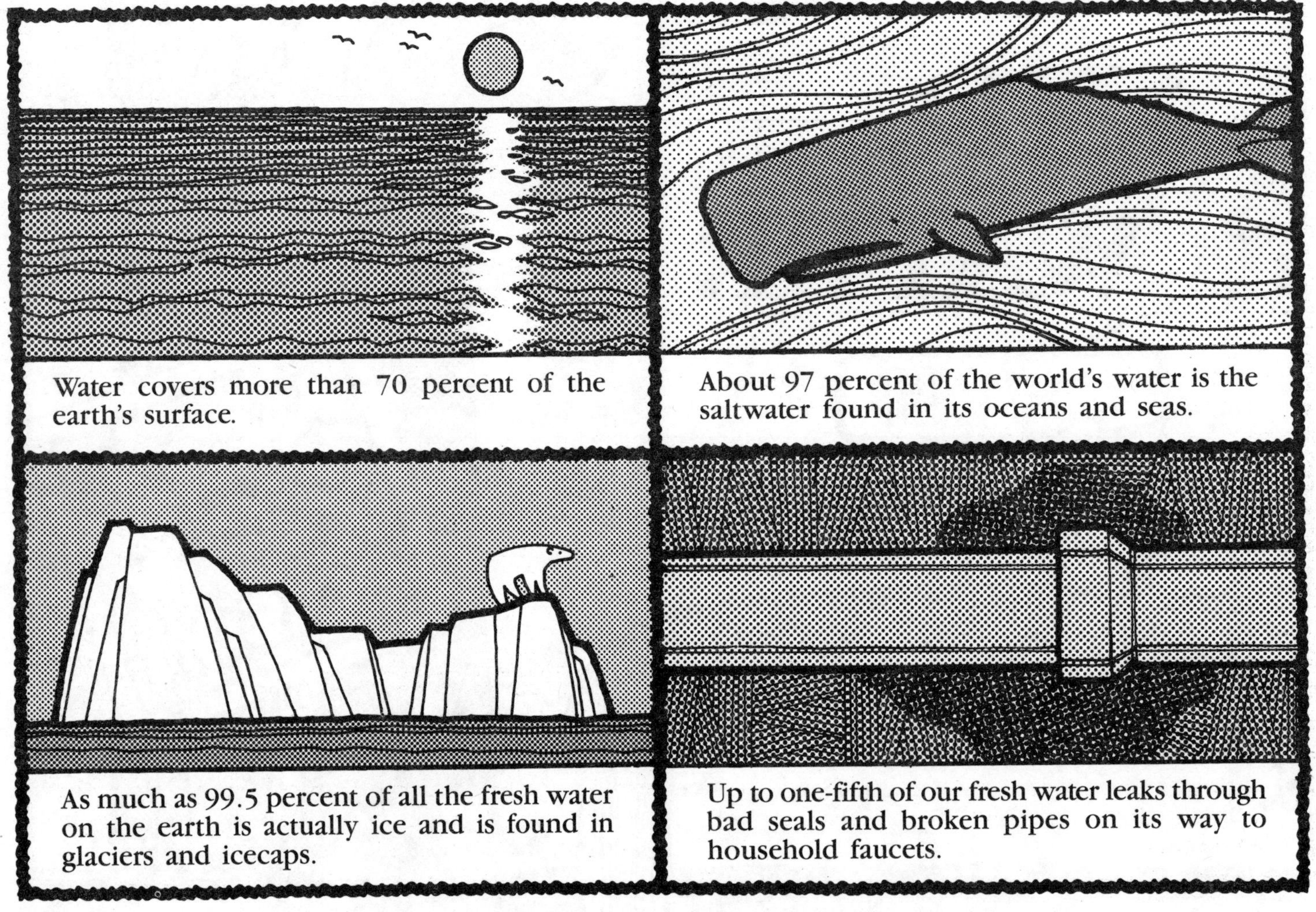

Air, land, water, and living things of all kinds are renewable resources.

Facts in Brief
(continued)

Clean water and fresh air are essential. A person can live only a few minutes without air and only a few days without water.

Copsa Mica, Romania, is called the "black town" because even its trees and grass have been blackened by the 30,000 tons of soot spewed out annually by two local factories. If horses remain in this town for more than a couple of years, they die from breathing its polluted air.

Almost 9 percent of East Germany's once-rich farmland has been ruined by fertilizers and pollutants.

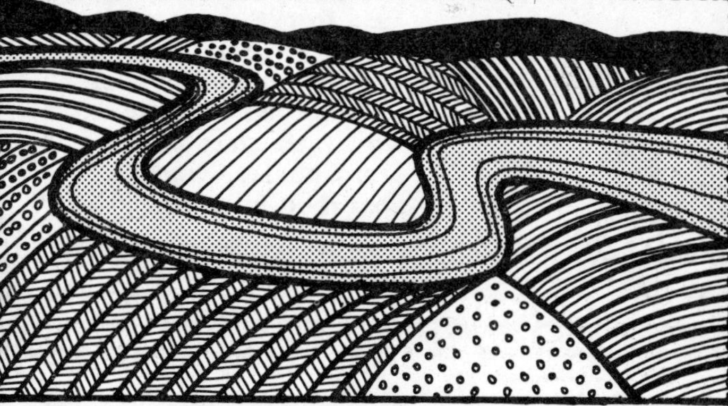

Pollution has made 95 percent of the river water in Poland unfit to drink.

Humanity faces two very old problems, living with itself and living with its natural surroundings.

The Water Cycle

Water does not stay in one place. Instead, it flows in a continuous cycle. Here are the phases of the water cycle.

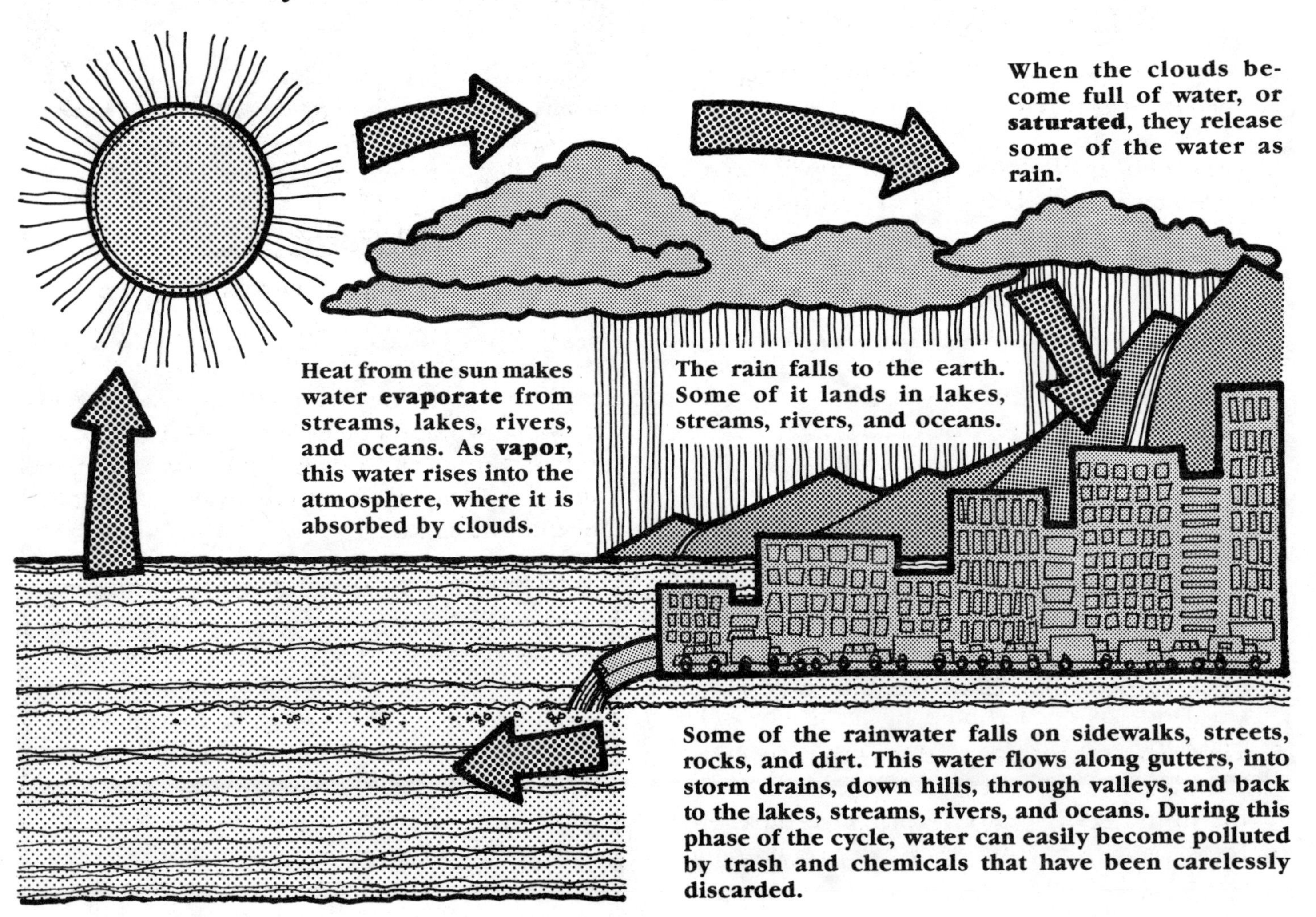

Watch the Water Cycle

Here's your chance to watch the water cycle.

WHAT YOU NEED

- ☐ a large metal or plastic salad or mixing bowl
- ☐ a hose or pitcher
- ☐ some water
- ☐ a sheet of clear plastic wrap (If possible, reuse a sheet that has been rinsed and dried.)
- ☐ a ceramic mug or other heavy, tall-sided container (If this container is opaque or colored rather than clear, it may be easier for you to see the water droplets in it.)
- ☐ a piece of string long enough to be tied around the bowl

WHAT YOU DO

1. Carry the bowl outside and place it in the sun.
2. Using a hose or pitcher, pour water into the bowl until it is about one-fourth full.
3. Gently place the mug in the center of the bowl, being careful *not* to splash any water into it.
4. Cover the top of the bowl with the clear plastic wrap.
5. Tie a piece of string around the top of the bowl to hold the plastic wrap in place and to make the inside of the bowl as airtight as possible.
6. Let the bowl sit in the sun while you watch to see what happens.

WHAT YOU WILL SEE

Heat from the sun will cause the water in the bowl to evaporate. This water will rise as vapor and condense on the plastic wrap, clouding it. As more and more water vapor condenses on the plastic wrap, it will form droplets. Then, these droplets will fall as "rain" back into the water in the bowl and into the mug.

VARIATION. Once vapor has condensed on the plastic sheet and made it look cloudy, gently move the bowl from the sun into the shade, being careful not to splash water onto the plastic sheet or into the mug. Let the bowl sit for a while. Then take the plastic sheet off and look inside. What happened to the water on the plastic sheet? Where did it go? Why?

Sources of Water

Water is essential to all living things. Plants, animals, and people need it to survive.

People of long ago relied on animals to help them find sources of fresh water. To be near water, these people settled along rivers and beside lakes. They drank the water and bathed in it. As time went on, they learned to use water in ways that made life more comfortable and more fun.

People who did not live near a river or a lake got water by digging wells. A **well** is a hole made in the ground to reach a natural deposit of liquid. In this instance, the liquid was **groundwater**, the water that lies under the ground in many places.

People also got water from falling rain and from melting snow. But rain and snow do not always fall when they are wanted or where they are needed. For this reason, people have had to find ways to move water and ways to store it.

Structures used to store water for future use are called **reservoirs**. Structures built especially to move water are called **aqueducts**. This word comes from two Latin words—***aqua***, meaning "water," and ***ducere***, meaning "to lead." Thus, an aqueduct leads, or channels, water from where it is to where it is needed.

EARTHWORDS

Let the fields and the gliding streams in the valleys delight me. Inglorious, let me court the rivers and forests.

—Virgil
Georgics I, line 485

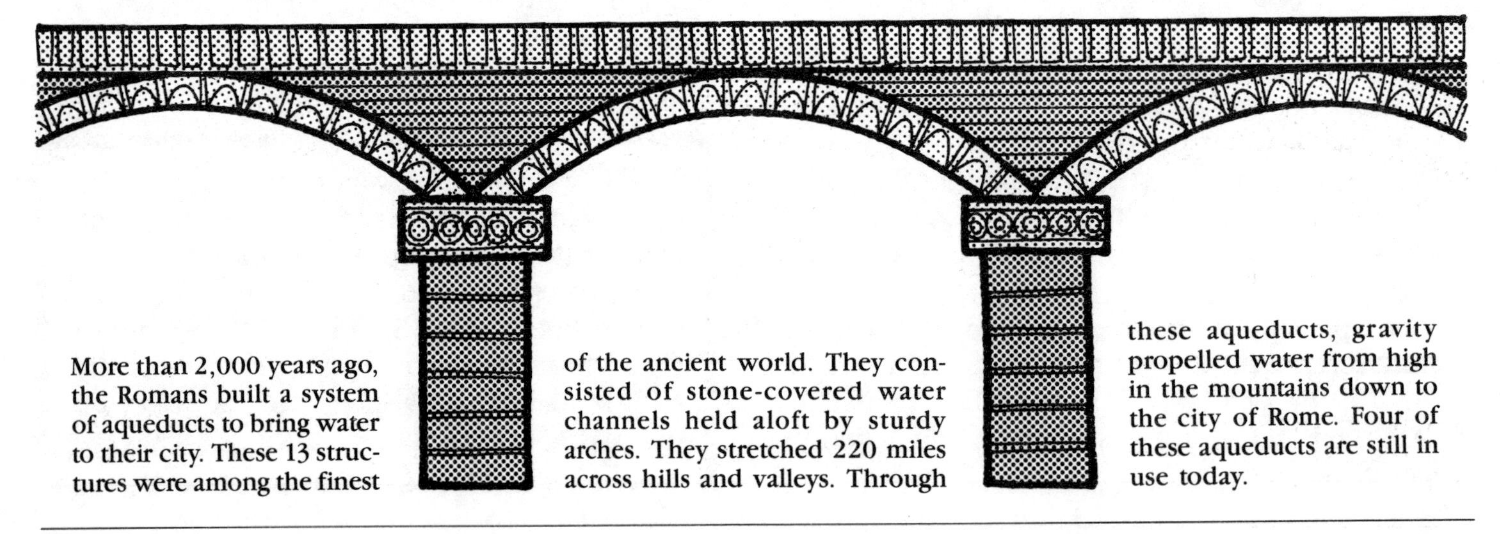

More than 2,000 years ago, the Romans built a system of aqueducts to bring water to their city. These 13 structures were among the finest of the ancient world. They consisted of stone-covered water channels held aloft by sturdy arches. They stretched 220 miles across hills and valleys. Through these aqueducts, gravity propelled water from high in the mountains down to the city of Rome. Four of these aqueducts are still in use today.

Sources of Water
(continued)

Today, most people in the United States get their water from metropolitan water systems. These systems may rely on one or several sources. For example, they may dam up rivers and create reservoirs to store runoff water for future use. They may tap natural springs or dig wells to get water from deep under the ground. They may also build aqueducts and lay pipe to carry water from nearby rivers and other places where water is plentiful to parched cities, where it is not.

First, find each of the cities listed in the chart below on a map. Then, do some research to add three more names to this chart.

HOW SOME MAJOR CITIES GET WATER

Use Surface Water	Use Underground Water	Pipe In Water
Atlanta	Honolulu	Denver
Chicago	Miami	Los Angeles
Dallas	San Antonio	New York

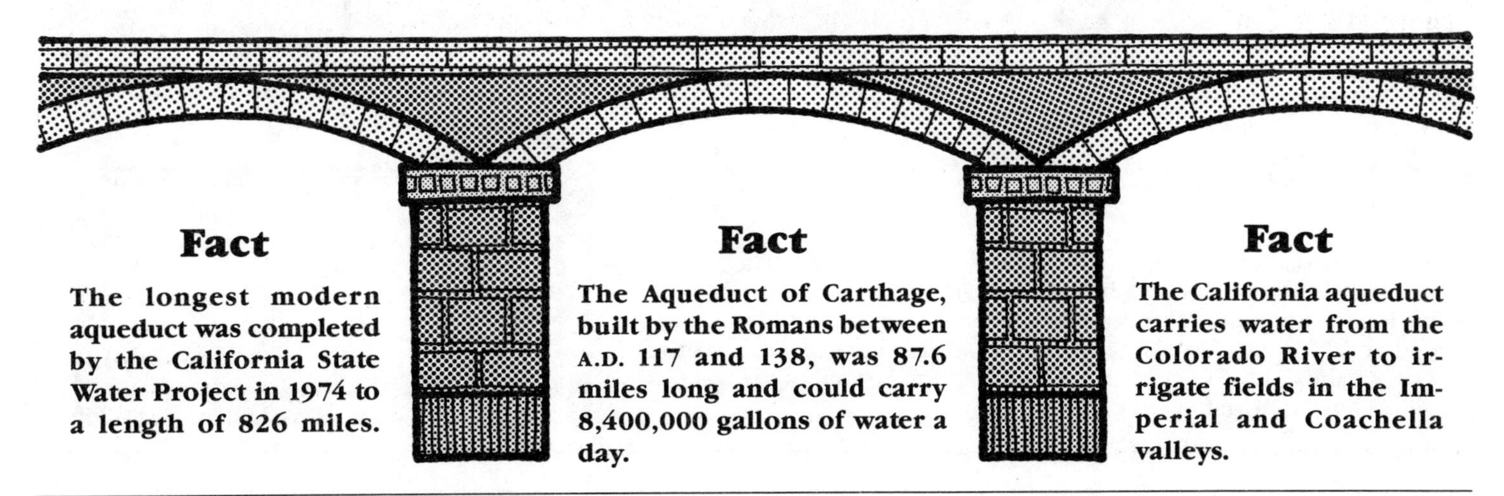

Fact

The longest modern aqueduct was completed by the California State Water Project in 1974 to a length of 826 miles.

Fact

The Aqueduct of Carthage, built by the Romans between A.D. 117 and 138, was 87.6 miles long and could carry 8,400,000 gallons of water a day.

Fact

The California aqueduct carries water from the Colorado River to irrigate fields in the Imperial and Coachella valleys.

Treatment of Water

Neither surface water in lakes and rivers nor underground water is pure. Both nature and people deposit substances in this water. Sometimes tiny organisms grow in the water. They carry diseases and can make people sick. For these reasons, water must be treated to make it safe for people to wash with, bathe in, and drink.

What happens at a water treatment plant?

Step 1. Intake

Large items, such as sticks, logs, fish, and plants, are screened out of water drawn from rivers, lakes, and other surface sources.

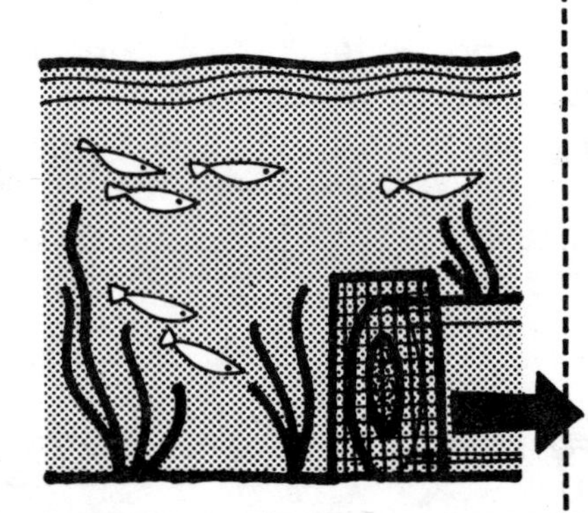

Step 2. Pretreatment

Chemicals, such as chlorine, alum, or lime, are added to remove impurities and to destroy unpleasant odors and tastes. Chemicals are also added to soften hard water by removing excess minerals. Then the water is mixed rapidly to distribute the chemicals evenly.

3. Coagulation

The water is allowed to collect and stand in a large basin. The chemicals that were added during pretreatment cling to any impurities in the water. This process in which chemicals and impurities gather together to form larger particles is called **coagulation**. As the water stands, these larger particles settle to the bottom of the basin and can be removed.

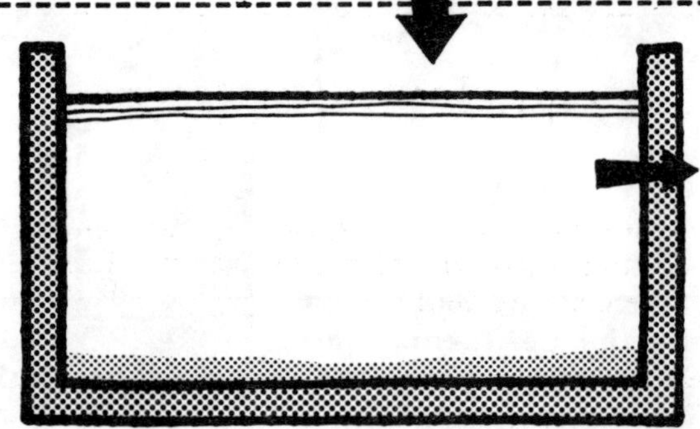

Treatment of Water
(continued)

Fact	Fact	Fact
• **Waterborne disease is one of the world's most serious health threats.**	• **In 1892, polluted drinking water from the Elbe River caused a cholera epidemic in Hamburg, Germany.**	• **In 1902, Belgium became the first country to use chlorine to kill bacteria in its water system.**

Step 4. Filtration

The water is filtered through layers of sand, gravel, and hard coal (called **anthracite**) to remove remaining impurities. An additional filter may be used to remove toxic organisms.

Step 5. Chlorination

Chlorine is added to the water to prevent the formation of bacteria. The chlorine is carefully measured so that the smallest effective amount is used. Sometimes, fluoride is also added to help prevent tooth decay.

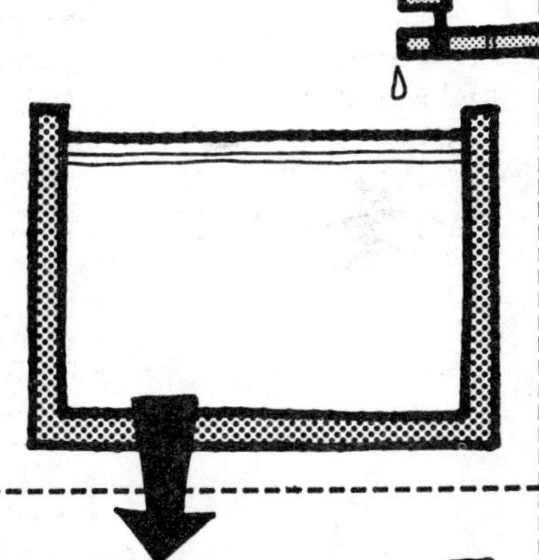

Step 6. Distribution

Now pure and ready for use, the water may be stored in a **reservoir** or tank. It may travel in pipes called **mains** to the homes or factories where it is needed.

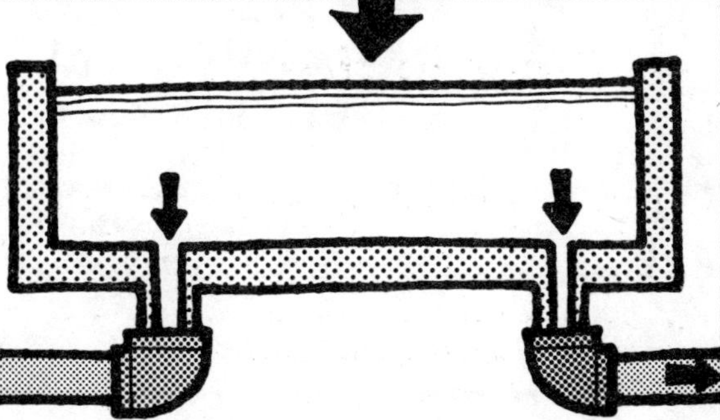

Distribution of Water

1. To the Water Treatment Plant

Both surface water from lakes and rivers and groundwater go to a water treatment plant (see pages 74–75).

2. By Mains and Service Lines to Water Users

The treated water travels by distribution pipes, or mains, and by service lines to water users, such as homes, factories, schools, hospitals, and businesses.

3. Through the Sewer System

Used water travels through the sanitary sewer system underground to the sewage treatment plant.

4. To the Sewage Treatment Plant

First, the water is cleaned. Next, it is carried to streams, where nature cleanses it more. Then, it is captured so that it can be used again.

5. Into Tanks, Open Reservoirs, and Covered Reservoirs

The clean, captured water is stored in large, elevated water tanks, in open reservoirs, or in covered reservoirs.

6. By Mains and Service Lines Back to Users

The water is either pumped electrically or gravity fed through mains to service lines and back to users.

Faucet Fun

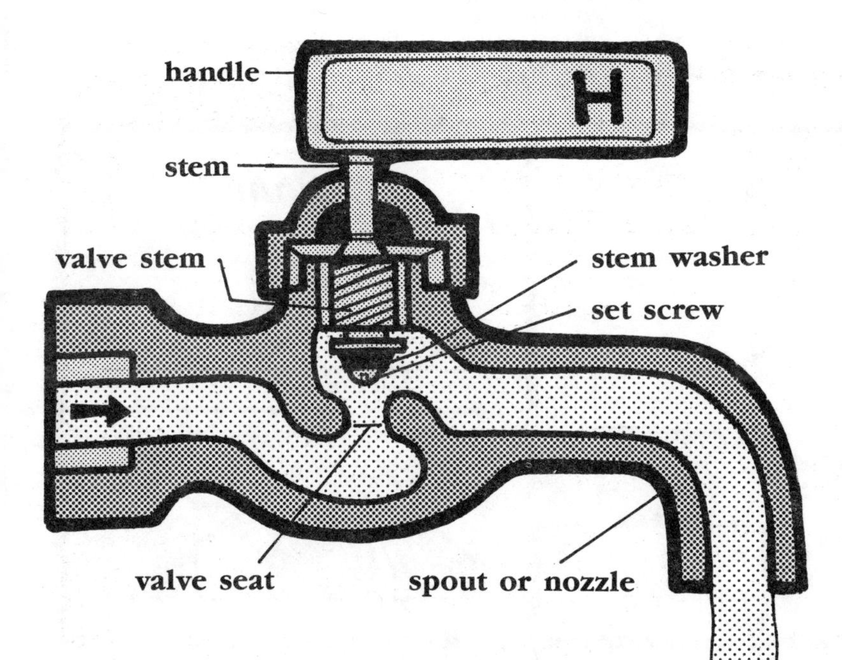

- Count and list all of the faucets inside your apartment or house. Look in bathtubs and shower stalls, not just at sinks and basins. And remember to include any faucets in a laundry room, workshop, and/or garage.

- Count and list all of the faucets around the outside of your apartment or house. Include all hose hookups and any special outlets that have been installed for fountains, spas, or swimming pools.

- Check all of these faucets regularly to be certain that none of them is leaking. If you find a leak, ask your parents to fix it or have it fixed. A small drip from a leaking faucet can waste more than 50 gallons of water a day! (See page 87.)

A **faucet** is one kind of valve. A **valve** is a device for controlling the flow of water from a pipe. It controls the flow by opening and closing part of the pipe. When the **faucet handle** is turned in one direction, a disk-shaped **washer** is raised away from the **valve seat** and the pipe is opened. Water from the pipe flows out the **spout**, or **nozzle**. When the handle is turned in the other direction, the washer is lowered firmly against the valve seat. The pipe is closed, and water does not flow out the spout.

Water Uses

We use water in many ways. For example, we use it for

Water Uses
(continued)

We also use water for

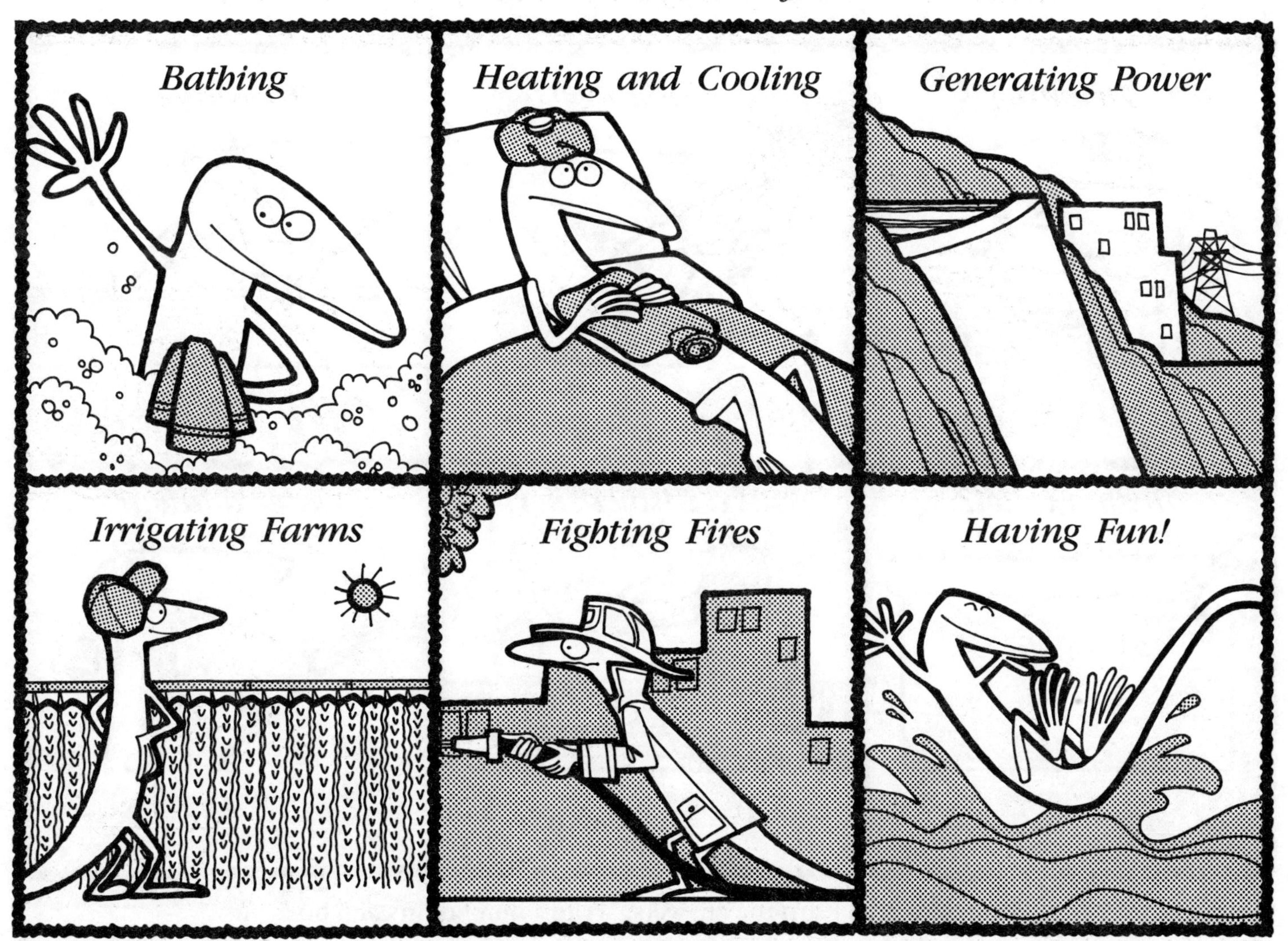

How Much Water?

How many gallons of water does it take to

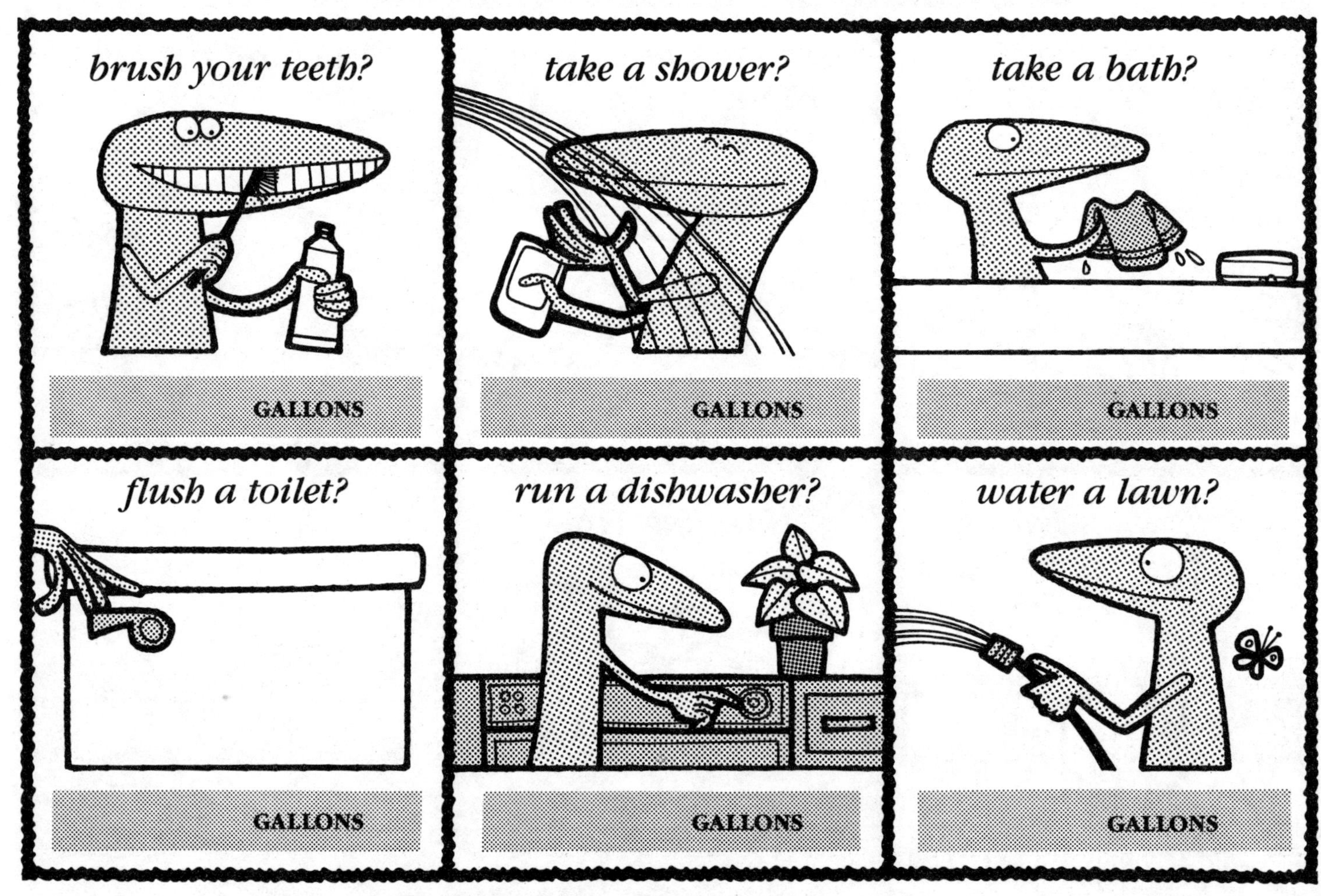

Do research to learn the answers. Write a number in each box.

Be a Water Detective

Contact your community's water department and ask these questions or others you've wondered about.

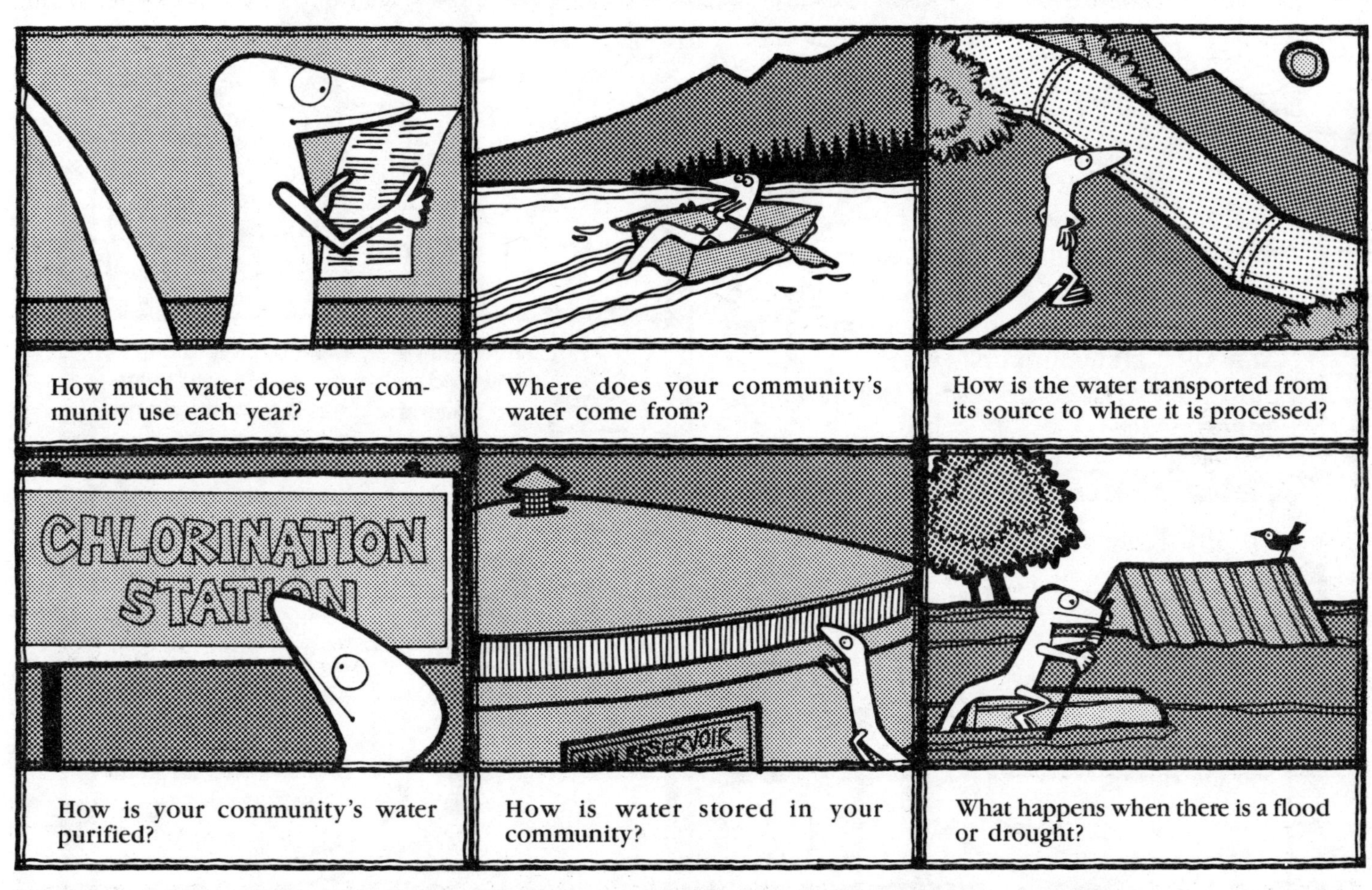

How is your family charged for the water it uses?

Home Water-Saving Ideas

Wash only full loads in your clothes washer. A washing machine can use up to 50 gallons of water per load.
taking baths and showers
cooking
washing dishes
washing clothes
flushing toilets
Ways we use water in our homes
When you water the lawn, make certain the moisture falls where it is needed. Don't water the sidewalk and the street. Instead of hosing down your driveway or patio, sweep away the dirt with a broom.
When you wash the family car, don't leave the water running. Instead, use a hose with a shutoff nozzle or fill a bucket with water and use that.

Shorter Showers

Facts

- **A two-minute shower uses about 24 gallons of water.**
- **A ten-minute shower uses more than 100 gallons of water.**
- **An average full-tub bath uses more than 40 gallons of water.**

WHAT YOU CAN DO

1. Take short showers instead of long showers or full baths.
2. Use a three-minute egg timer and try to complete your shower before the time is up.
3. Turn off the water while you shampoo your hair or lather up your skin and then turn it on when you are ready to rinse off the suds.
4. Ask your mom or dad to install low-flow shower heads. These devices slow the water flow from 12 to 3 gallons a minute and reduce shower water consumption by 75 percent.

Shortest Shower Award

Make a chart like the one shown below. Write the names of the members of your family in the column on the left. For one week, have each family member time his or her showers and record the number of minutes each shower takes.

At the end of the week, total the minutes each family member has spent showering. Divide this total by the number of showers that family member took during the week to determine the **average** number of minutes spent in the shower.

Create a **Shortest Shower Award** and present it to the person in your family who has averaged the fewest minutes per shower.

This
Shortest Shower Award
is presented to

____
(name)

for conserving water by averaging
only ______ minutes per shower
(number)
between __________ and __________.
(date) (date)

NOTE: Any family member who goes without taking a shower for the entire week is both dirty and disqualified. To be eligible for the award, a family member must take at least three showers!

Name of Family Member	Minutes Spent Showering On							Sum of This Week's Showers
	Mon.	Tue.	Wed.	Thur.	Fri.	Sat.	Sun.	

Chart a Flush

You use 6 gallons of water each time you flush a toilet.

1. Keep track of the number of times you flush a toilet each day for one week.

Monday	Tuesday	Wednesday	Thursday	Friday	Saturday	Sunday

2. To find out how many times you flushed a toilet during the week, add your daily totals. Record the sum on the line below.
3. To discover how many gallons of water you flushed away in one week, multiply the total number of flushes by 6. Record the product on the line below.

______ **flushes** × 6 gallons per flush = ______ **gallons**

(Total Number of Flushes) **(Water Flushed Away)**

4. During the next week, try flushing the toilet less often. You may not need to flush it every time you use it. Again, keep track of the number of times you flush each day.

Monday	Tuesday	Wednesday	Thursday	Friday	Saturday	Sunday

5. Subtract the number of gallons you flushed away during the second week from the number you flushed away during the first week to see how many gallons of water you saved.

______ **gallons** − ______ **gallons** = ______ **gallons**

(Flushed Away in First Week) **(Flushed Away in Second Week)** **(Saved by Flushing Less Often)**

Leaky Faucets

Discover for yourself how much water is wasted by one dripping sink or bathtub faucet.

Liquid Measure Equivalents

2 cups = 1 pint
4 cups = 2 pints = 1 quart
16 cups = 8 pints = 4 quarts = 1 gallon

WHAT YOU NEED

- ☐ one leaky faucet
- ☐ a watch or clock
- ☐ a piece of paper
- ☐ a pencil or pen
- ☐ some patience
- ☐ a liquid measuring cup
- ☐ an empty bucket

WHAT YOU DO

1. Find a leaky sink or bathtub faucet.
2. Close the drain below this faucet.
3. Look at a watch or clock and write the exact time on your piece of paper.
4. Be patient and let dripping water accumulate in the sink or tub.
5. Check the rising water level frequently so you won't have an overflow.
6. At the end of 12 hours, measure the water that has accumulated. Use the measuring cup to empty the dripped water from the sink or tub into the bucket. Each time, fill the measuring cup as full as possible. Count the number of times you fill the cup, and write this number on your piece of paper.
7. To find out how much water leaked from your dripping faucet in 12 hours, multiply the capacity of the measuring cup by the number of times you filled it.
8. Don't waste the water in the bucket. Instead, empty it onto a flower bed or lawn, or use it to wash the car.
9. Talk to your parents about getting the leaky faucet fixed. Probably all that is needed it a hard round rubber piece called a **washer**. A new washer will cost less than fifty cents.

Fact

- **Even a small drip from a leaky faucet can waste more than 50 gallons of water a day.**

Fact

- **50 gallons of water is a lot to waste! It's enough water to flush a toilet eight times.**

Fact

- **50 gallons is enough water to run an automatic dishwasher twice on full cycle or to wash the dishes from six meals by hand.**

Phosphate Problems

A **detergent** is a cleansing agent. Most detergents are liquids or powders that will dissolve in water to become liquids. Detergents are used to wash dishes and clothes. They clean by making fibers soak up water and by putting oil and dirt into suspension so that both can be rinsed away.

Manufacturers add chemicals called **phosphates** to detergents to increase the amount of suds they produce and to make these suds last longer. But phosphates have other effects. For example, they encourage plant growth by enabling plant leaves to make food. While small amounts of phosphate are good for the environment, large amounts are harmful to it.

Phosphates biodegrade slowly, and their effects are felt for a long time. Phosphates find their way into streams and lakes where they increase the acidity of the water. They also speed up the growth of plants called algae. Algae block out light and choke off water flow. Algae rob other plants of the fresh water and nourishment they need to survive. When these plants die, decaying plant material pollutes the water and poisons the fish and other animals that live in it or drink from it.

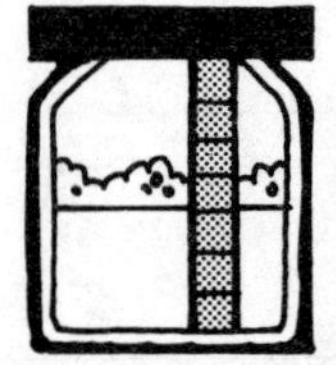

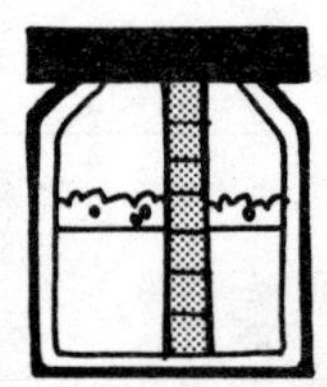

This experiment will give you an opportunity to observe the relationship between phosphates and sudsing.

WHAT YOU NEED

- ☐ one tablespoon for measuring
- ☐ four different brands of detergent
- ☐ four identical pint jars with lids
- ☐ a roll of masking tape
- ☐ a pair of scissors
- ☐ a ruler
- ☐ a ballpoint pen
- ☐ some water
- ☐ a watch, clock, and/or kitchen timer

WHAT YOU DO

1. Run a strip of masking tape from top to bottom down the side of each jar.
2. Using the pen and ruler, mark off the tape in inches. Then insert additional marks to divide the inches into halves and fourths.
3. Measure one tablespoon of detergent into each jar.
4. As you do so, write the name of the detergent on a strip of masking tape and attach the tape strip to the side of the jar.

Phosphate Problems
(continued)

5. Fill each jar half full of water.
6. Mark the actual water line on the tape.
7. Screw the lids on the jars tightly.
8. Shake the first jar for one minute.
9. When the minute is up, mark a line on the tape to show the height of the suds.
10. Repeat steps 8 and 9 for the detergent samples contained in the remaining three jars.
11. Allow the jars to stand undisturbed.
12. After 10 minutes, mark the level of suds in each jar.
13. After another 10 minutes, again mark the level of suds in each jar.
14. Summarize your observations on a chart like the one shown here.

BONUS PROJECT. Illustrate the relationship between the phosphate level and the suds height by using the data from your chart to create a line or bar graph.

Phosphate Chart

Brand Name of Detergent	Percent of Phosphate as Shown On Product Label	Height of Suds After 1 Minute	Height of Suds After 10 Minutes	Height of Suds After 20 Minutes
1.				
2.				
3.				
4.				

What Is Acid Rain?

To understand acid rain, you need to know something about acids. **Acids** are water-soluble chemical compounds that have several very specific characteristics. For example, acids redden litmus paper. They make your eyes sting and your skin burn, and they taste sour. Some acids are so strong, or **caustic**, that they can actually pit metal and dissolve rocks.

Lemon juice and vinegar are common household acids. Lemon juice tastes sour. If you accidentally get some in your eye, it will sting. Normally, lemon juice won't make your skin burn unless it is already open or irritated. Even though lemon juice is a relatively mild acid, it can stain or even pit the coating on a kitchen sink or counter within a short time if allowed to sit rather than being wiped away promptly.

Chemicals called **sulfur dioxide** and **nitrogen dioxide** are spewed into the air from industrial chimneys and automobile exhaust pipes. When these airborne chemicals become dissolved in rainwater, the result is weak, or **dilute**, **sulfuric acid** or **nitric acid**. These acids fall to earth as **acid rain**.

Acid rain damages many man-made structures. Limestone and marble are rocks used to create buildings, bridges, gravestones, and statues. These rocks react with sulfuric acid by dissolving. After many seasons of acid rain, these structures can become pitted, change shape, or be weakened.

Acid rain also harms the environment in other ways. When acid rain soaks into the ground, it dissolves valuable minerals and carries them away. Acid rain burns leaves, slows plant growth, and changes the chemical characteristics of the streams and lakes into which it falls. Acid rain may affect the food supplies for fish and even prevent their eggs from hatching.

Fact

The basic fuel of Eastern Europe is a high-sulfur coal called **lignite**.

Fact

Along the border between East Germany and Czechoslovakia, acid rain has stripped entire pine forests of their needles.

Fact

Acid rain has damaged as much as 80 percent of the lush forests that once flourished in East Germany.

An Acid Rain Experiment

How acid is your rainwater? Do this experiment to find out.

WHAT YOU NEED

- ☐ a clean widemouthed jar (A 22-ounce plastic peanut butter jar is ideal.)
- ☐ litmus paper and color chart (available from stores that sell pets and/or tropical fish)
- ☐ a rainy day

WHAT YOU DO

1. Set the jar out in the open—away from eaves, gutters, and trees—where rainwater will fall directly into it.
2. Wait patiently while rainwater collects in the jar.
3. When the rain stops, dip a strip of litmus paper into the collected rainwater.
4. Determine whether rainwater in your area is acid, alkaline, or neither ("pH balanced") by comparing the color of the dampened litmus paper to the colors shown on the chart.

EARTHWORDS

Probable nor'east to sou'west winds, varying to the southard and westard and eastard and points between; high and low barometer, sweeping round from place to place; probable areas of rain, snow, hail, and drought, succeeded or preceded by earthquakes with thunder and lightning.

—Mark Twain
New England Weather

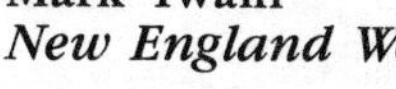

Is It an Acid?

Use this simple test to measure the acidity of a variety of common household liquids.

WHAT YOU NEED

- ☐ litmus paper and color chart (available from stores that sell pets and/or tropical fish)
- ☐ several common household liquids, such as dishwashing or laundry detergent, lemon juice, milk, tap water, and/or tomato juice
- ☐ enough clean widemouthed containers (such as cups or margarine tubs) to hold small amounts of these liquids
- ☐ a piece of paper
- ☐ a pencil or pen

WHAT YOU DO

1. Pour a small amount of each liquid you plan to test into a separate container.
2. If the dishwashing or laundry detergent is a powder or a very thick liquid, mix it with a small amount of tap water.
3. Dip a strip of litmus paper into each liquid. Use a new strip each time.
4. Determine which liquids are acids and which ones are **bases** by comparing the colors of the dampened litmus paper strips to the colors shown on the chart.
5. Record your results by noting which liquids are acid, which are alkaline, and to what degree.
6. Summarize your results on a chart or graph.

EARTHWORDS

Autumn came. The clouded sky descended low upon the black contours of the hills; and the dead leaves danced in spiral whirls under naked trees, till the wind, sighing profoundly, laid them to rest in the hollows of bare valleys. And from morning till night one could see all over the land black denuded boughs, the boughs gnarled and twisted as if contorted with pain, swaying sadly between the wet clouds and the soaked earth.

—Joseph Conrad
Tales of Unrest

How Clean Is the Water?

Water can be made unclean, or impure, by things you cannot see. For example, bacteria and other tiny organisms can grow in water and make it unsafe to drink. Water can also be made dirty by things you can see but may not notice. How clean is the water? Test some samples of it to find out.

WHAT YOU NEED

- ☐ four small clean empty containers (Baby food jars are ideal.)
- ☐ four coffee filters
- ☐ some masking tape
- ☐ a pencil or ballpoint pen
- ☐ a small notebook for your observations, thoughts, and questions
- ☐ a large widemouthed jar

WHAT YOU DO

1. In the small containers, collect samples of water from four different sources. For example, it might be fun to compare faucet water, bottled water, rainwater, and pond water.
2. Using the masking tape and a pencil or pen, carefully number and label each water sample as you take it.
3. On separate pages in your notebook, identify each sample by source and write any extra information about this sample that may help you interpret your results. For example, you might wish to note the time of day when you collected the sample, the type or amount of rainfall, the location of the pond, and whether the water in the pond appeared to be clear or cloudy. You might also wish to note whether any of your samples had an odor.
4. Hold a coffee filter in the mouth of the large jar and gently pour the water from one of your samples through this filter into the jar.
5. Look closely at the filter. To what degree was it discolored by the water you poured through it? Were many particles caught by the filter? Can you identify any of these particles? In your notebook, make some notes about this water sample and what you observed.
6. Repeat steps 4 and 5 for the other three water samples, using a new filter each time.

THINGS TO THINK ABOUT

- Which water sample left behind the most color and/or residue?
- Which water sample left behind the least color and/or residue?
- What might account for these differences?

All About Oil

Oil is a natural product found in rock pockets inside the earth. It is called a **fossil fuel** because it was created millions of years ago by the decomposing remains of plants and animals, and it can be burned to release energy. It is also called **petroleum**. This word comes from two Latin words, ***petra*** meaning "rock" and ***oleum*** meaning "oil." The Romans thought of petroleum as "rock oil."

Since oil was first discovered, it has proved to be very useful. Ancient people used it to seal the rock walls of their homes and to waterproof the hulls of their wooden boats. American Indians used it to make medicine and paint. Today, oil is used to make dyes, fuels, lubricants, medicines, paints, and plastics.

In some places, petroleum oozes to the surface of the earth or floats atop pools of water. In most places, it lies hidden beneath the ground. Scientists known as **geologists** study rock formations. They draw maps to show the places where oil

Fact

- Because we have no way to make new oil to replace what we use, oil is called a **nonrenewable resource.**

Fact

- The modern oil industry began in 1859, when a producing well was drilled in Pennsylvania at a place that later became Titusville.

Fact

- While tanker spills do make headlines, more than half of the oil that pollutes our oceans comes from land-based sources.

All About Oil
(continued)

might be found. Petroleum companies lease land and drill wells in these places.

Drilling for oil is hard work. First, oil field workers, called **roustabouts**, erect a tall structure called a **derrick**. The derrick supports the platform on which the roustabouts work and houses some of their equipment. Next, the workers use sharp **bits** that turn round and round to cut through loose dirt, heavy clay, sandstone, and solid rock. Every now and then, they take **core samples** and examine them for signs of "black gold."

Drilling for oil can also be dangerous. Sometimes natural gas trapped underground causes wells to "blow out," or explode with tremendous force. Oil shoots high into the air, and workers must struggle to control this **gusher**. At other times, oil wells catch fire and burn for days. Because oil fires cannot be put out with streams of water, experts must be called in to fight them.

When the workers finally reach oil, they extend pipes down to it. Then, they use a big pump to pull the oil through these pipes from beneath the ground to the surface. Pipes, trucks, and ships carry this crude oil from the fields where it was found to the cracking plants and refineries where it will be turned into many useful products.

Transporting oil from where it is found to where it is refined and used can be hazardous to the environment. Sometimes pipelines leak. Sometimes oil trucks are involved in accidents. Sometimes oil tankers tear their hulls on anchors or rocks and spill their contents into the sea.

What You Can Do

Oil pollution is not limited to accidents that happen on wells or while oil is being moved from place to place. In fact, one of the biggest causes of oil pollution is motor oil that is improperly discarded after its use. Talk with your parents about what they do with their used motor oil. Suggest that when they change the oil in their cars or trucks, they pour the used oil into a large container. When this container is nearly full, they can take it to a licensed gas station. Attendants at these stations will recycle the used oil or dispose of it in a way that is safe for the environment.

Cleaning Up Oil Spills

Over the past twenty-five years, crude oil has been accidentally spilled into the world's oceans hundreds of times. Some of these spills are small and do little damage, but other spills are large. Each forms a widespread film of oil atop the water. This floating film is called an **oil slick**.

The slicks created by large spills have a devastating effect on the ocean and on beaches and shores. The oil seeps into eggs laid at the water's edge and kills the tiny animals growing inside. The oil clogs the gills of fish, making it impossible for them to breathe. The oil coats the feathers of birds. In doing so, it eliminates the air layer that insulates them from the cold. It also reduces their buoyancy in water, leaving them unable to dive for food or to swim away from predators. The oil kills the marine animals who eat food or drink water poisoned by it.

A 1969 leak in an offshore well created an oil slick that stretched 60 miles along the California coast. During the first four months after that spill, 3,000 birds died. On March 24, 1989, the oil tanker *Exxon Valdez* ran aground near Valdez, Alaska. More than 10 million gallons of crude oil spilled from the tanker's torn hull into Alaska's Prince William Sound. As a result, at least 33,000 seabirds, 980 sea otters, and 136 bald eagles have died; and the toll continues to rise.

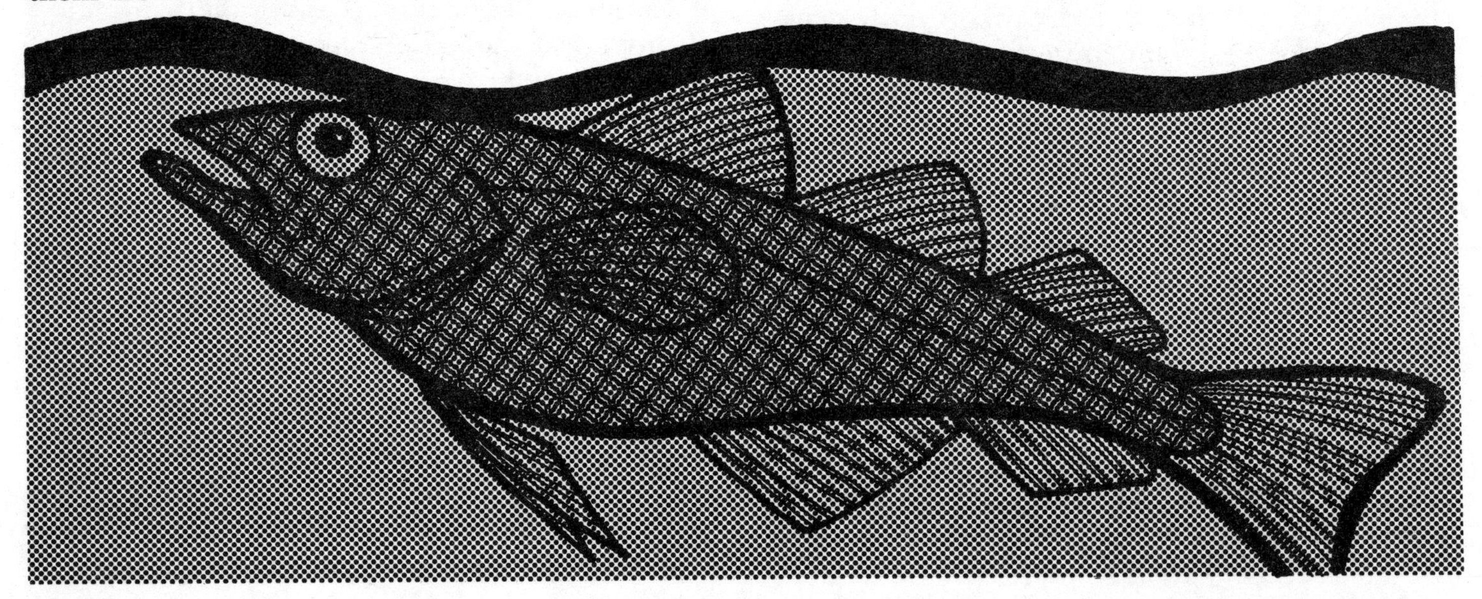

Cleaning Up Oil Spills
(continued)

To understand why spilled oil is difficult to remove from water, try this experiment.

WHAT YOU NEED

- ☐ a shallow, rectangular baking pan
- ☐ a small amount of cooking oil
- ☐ some cold water
- ☐ a variety of cleaning materials to be tested (For example, you might try baking soda, cotton balls, dishwashing liquid, laundry detergent, old nylon stockings, paper towels, shredded newspapers, and/or sponges.)

WHAT YOU DO

1. Pour water into the pan until it is half full.

2. Spill a small amount of oil into the water and watch as it forms a slick on the surface.

3. Try to clean up the spill using the cleaning materials you have selected. In separate sections of the pan, test the effectiveness of at least four different ones.

4. Evaluate the effectiveness of the materials you used. Assign each one a rating from 1 (least effective) to 5 (most effective). Which material worked best?

THINGS TO THINK ABOUT

- Would the material you found to be most effective in cleaning up a small oil spill in a pan be useful in cleaning up a large oil spill in an ocean? Why or why not?

- Did any of your cleaning materials leave behind a residue that might be damaging to the environment? If so, how would you clean up this residue?

- When a water-resistant coat or jacket is washed or cleaned improperly, it loses its water resistance. You may be unaware of this loss until you get thoroughly soaked during the next rainstorm. Sometimes a detergent used to clean oil-soaked birds removes the natural oil from their feathers, allowing them to become similarly water-logged. Might some of your cleaning materials have this or other harmful side effects?

Adopt a Stream

To have clean rivers and lakes, we must have clean streams; but streams can easily become polluted. Fertilizers and pesticides wash into streams from farmers' fields. Pipes carrying sewage break and leak wastes into streams. Dirt eroded from bare land clogs and chokes the streams into which it flows. Litter carelessly dropped beside a stream often falls or blows into the water.

Polluted stream water can make animals and people sick. It can even kill the fish and insects that live in the stream. These insects keep stream water clean by eating decaying plant and animal matter that might otherwise poison the stream. They help to break up human and animal wastes. They also serve as food for many of the fishes, birds, and amphibians that live in or near the water.

With a group of friends or classmates, adopt a stream in your community. Take steps to protect it from pollution. For example, you might pick up litter alongside the stream so that it will not fall into the water. You also might look for possible sources of pollution and report any you find to adults who can decide what should be done. For more information and ideas, write to

Save Our Streams
The Izaak Walton League of America
1401 Wilson Boulevard
Level B
Arlington, Virginia 22209

EARTHWORDS

Government cannot close its eyes to the pollution of waters, to the erosion of soil, to the slashing of forests any more than it can close its eyes to the need for slum clearance and schools.
—Franklin Delano Roosevelt

The Pesticide Problem

A **pesticide** is a poison used to kill pests. To a farmer, weeds are pests because they make it hard for young plants to grow. Insects and small rodents are also pests because they damage plants and destroy crops.

About 70 percent of America's farm crops are sprayed with pesticides. Over time, much of the poison disappears, but some of it stays on the plants. This leftover poison is called **pesticide residue**.

Pesticide residue is a serious problem. People eat it on fruits and vegetables. Farm animals eat it on grasses and grains. Pesticide residue that is eaten by animals can contaminate the meat, milk, and eggs that come from these animals.

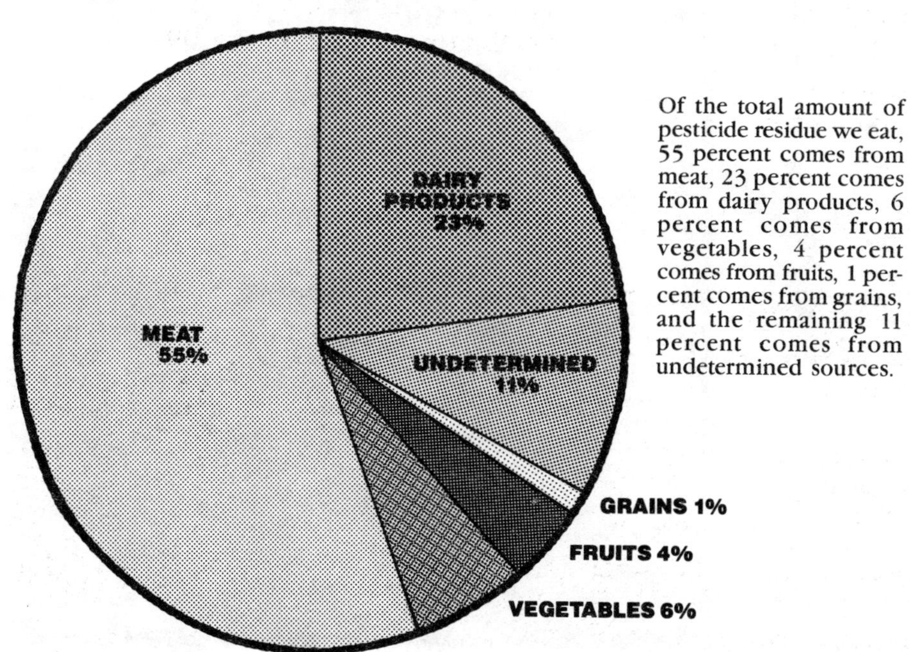

Of the total amount of pesticide residue we eat, 55 percent comes from meat, 23 percent comes from dairy products, 6 percent comes from vegetables, 4 percent comes from fruits, 1 percent comes from grains, and the remaining 11 percent comes from undetermined sources.

Rachel Carson

1907 – 1964

As a child, Rachel Carson loved nature and had a strong desire to write. In college, she studied both composition and biology. As a scientist, she became more and more concerned about pesticides. The strong chemicals used to kill harmful insects were also killing beneficial insects and many birds and fishes as well. Rachel feared that the result would be a bleak season in which there would be no bees to pollinate plants and no birds to sing. She described this season in a book called *Silent Spring*. Because of her book, laws were passed restricting the use of DDT and other pesticides in the United States and in many countries throughout the world.

Check Your Home for Toxics

A **toxic substance** is a chemical or mixture of chemicals whose manufacture, distribution, use, or disposal may present an "unreasonable risk" to the health of a person and/or the environment. Toxic substances—sometimes called **toxics** for short—can harm people if they are eaten, inhaled, or absorbed through the eyes or skin. The effects of toxic substances vary widely. They range from dizziness and vomiting to changes in behavior and mental alertness. Some toxics are known to cause cancer.

Household chemicals can be toxic if they are not used, stored, and/or disposed of properly. Most of the dangerous substances in your home are found in cleansers, pesticides and solvents. First, look around your house and garage for some of the products listed in the box and put a check mark beside each one you find. Next, read the warning labels on these products. Then, discuss with your mom or dad the best way to use these products safely. With your parents' help, find a place to store these potentially toxic products so they will be out of the reach of younger children in your family. Finally, discuss with your parents the proper ways to dispose of toxics.

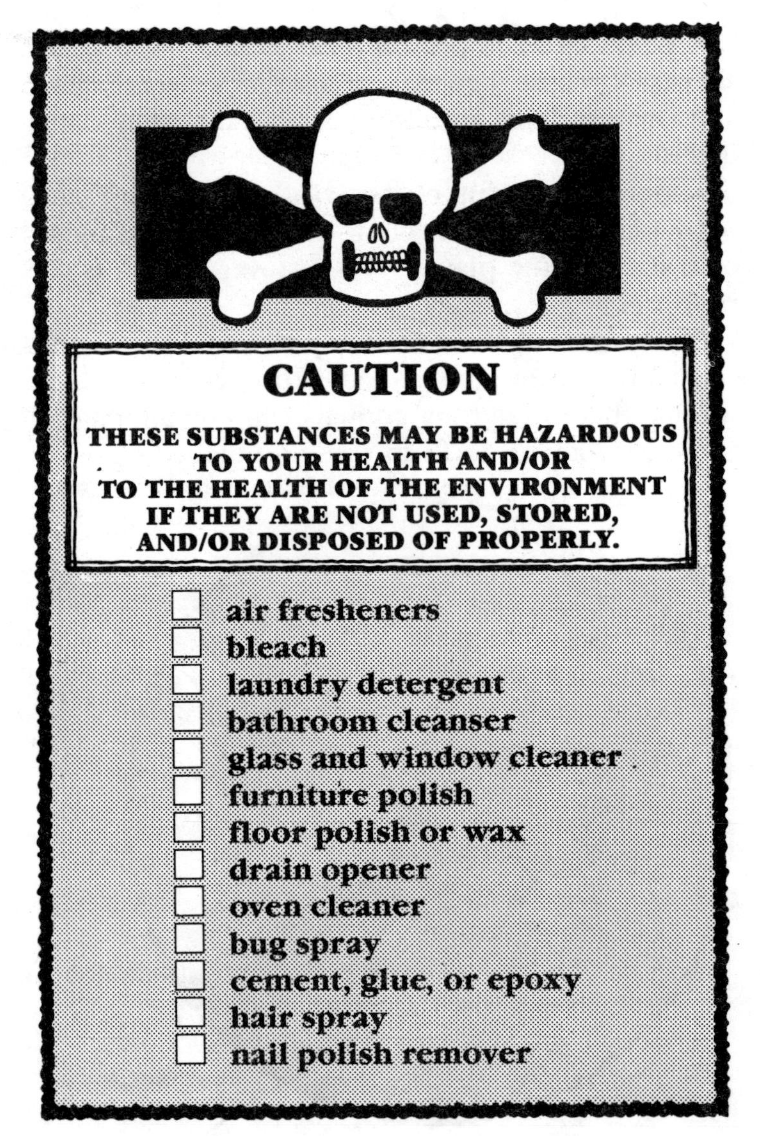

Nontoxic Alternatives

Pretend that you have been hired to create a nontoxic alternative for a potentially toxic product you and your family currently use. For example, you might make a mixture of olive oil and lemon juice to use instead of commercial furniture polish.

1. Decide what kind of product you want to create. Brainstorm to come up with several product ideas just in case some of them don't work out.
2. Collect all of the necessary ingredients in one convenient work area.
3. Carefully measure and mix the ingredients to create your product. As you do so, make detailed notes about the amount of each ingredient you use. These notes will become your recipe, or formula. If they are accurate, you will be able to make another batch of your product which is *exactly* like the first batch or you will be able to change the ingredient proportions to improve your product.
4. Test your product carefully. Does it do what you intended it to do? Does it do things that you did not intend? If so, are these things good or bad, advantages or disadvantages?
5. Make any necessary adjustments in the recipe, or formula, for your product and test it again.
6. When you are satisfied with your product, give it a name.
7. Design an earth-friendly package for your product. (For some packaging ideas, see page 28.)
8. Decide how much the ingredients for your product cost and how much you will need to charge for the product when you sell it.
9. Write a brief paragraph describing your product.
10. List the advantages of using your product instead of using similar ones that are already on the market.
11. Write a slogan or jingle to help sell your product.
12. Create a brochure or a full-page advertisement for your product using what you did in steps 6–11 above.

EARTHWORDS

There is a sufficiency in the world for man's need but not for man's greed.

—Mohandas Gandhi

Community Pollution Checklist

An **air pollutant** is anything that makes the air impure or dirty. We can see air pollutants like stone and wood dust or textile fibers. We cannot see other air pollutants like carbon monoxide, hydrocarbons, nitrogen dioxide, sulfur dioxide, and ozone. Air pollutants cause a variety of physical discomforts, including headaches, itchy eyes, and sore throats. The more caustic ones can also blister paint, erode stone, and etch metal.

Look for some of the following sources of air pollution in your community.

Test the Air

To check the air you breathe for visible pollutants, try this experiment.

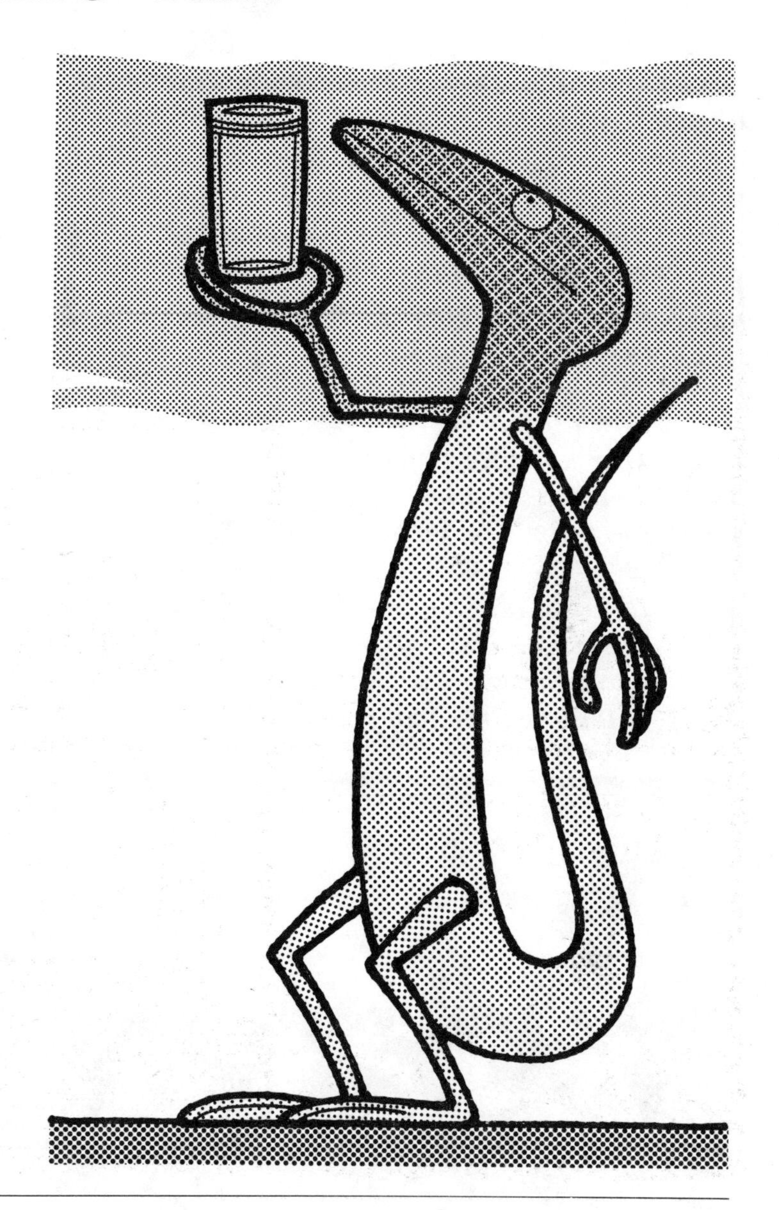

WHAT YOU NEED

- ☐ a clean widemouthed medium-sized glass jar
- ☐ some petroleum jelly

WHAT YOU DO

1. Spread a thin, even layer of petroleum jelly on the inside of the jar.

2. Take the jar outside and place it in the open air. Make certain that it is away from the sidewalk or playground and at least three feet off the ground so that dirt will not be kicked into it.

3. Leave the jar outside for five days.

4. Observe the results and answer these questions.

   a. Would you say that the petroleum jelly is only slightly dirty, medium dirty, or very dirty?

   b. Is most of the dirt you see textile fibers, stone dust, wood dust, or tire dust?

VARIATION. Conduct the same experiment in a different place. Compare the results. What factors might account for any significant differences you observe?

Walk!

Walk!

Next time you have to go someplace
That's fairly close to your home,
Instead of riding with your folks,
Just get there on your own.

Cars pollute the air we breathe—
They use a lot of fuel—
So ride your bike or walk instead
Next time you go to school.

It's a simple step and healthful,
And you help to do your share;
For when you walk or ride your bike,
You show the earth you care.

For one week, travel to school in a way that uses only *your* energy and will *not* pollute the earth. For example, you might pedal on a bicycle or a unicycle, glide on a scooter, roller skate, hop on a pogo stick, or walk. Use your imagination, travel safely, and have fun!

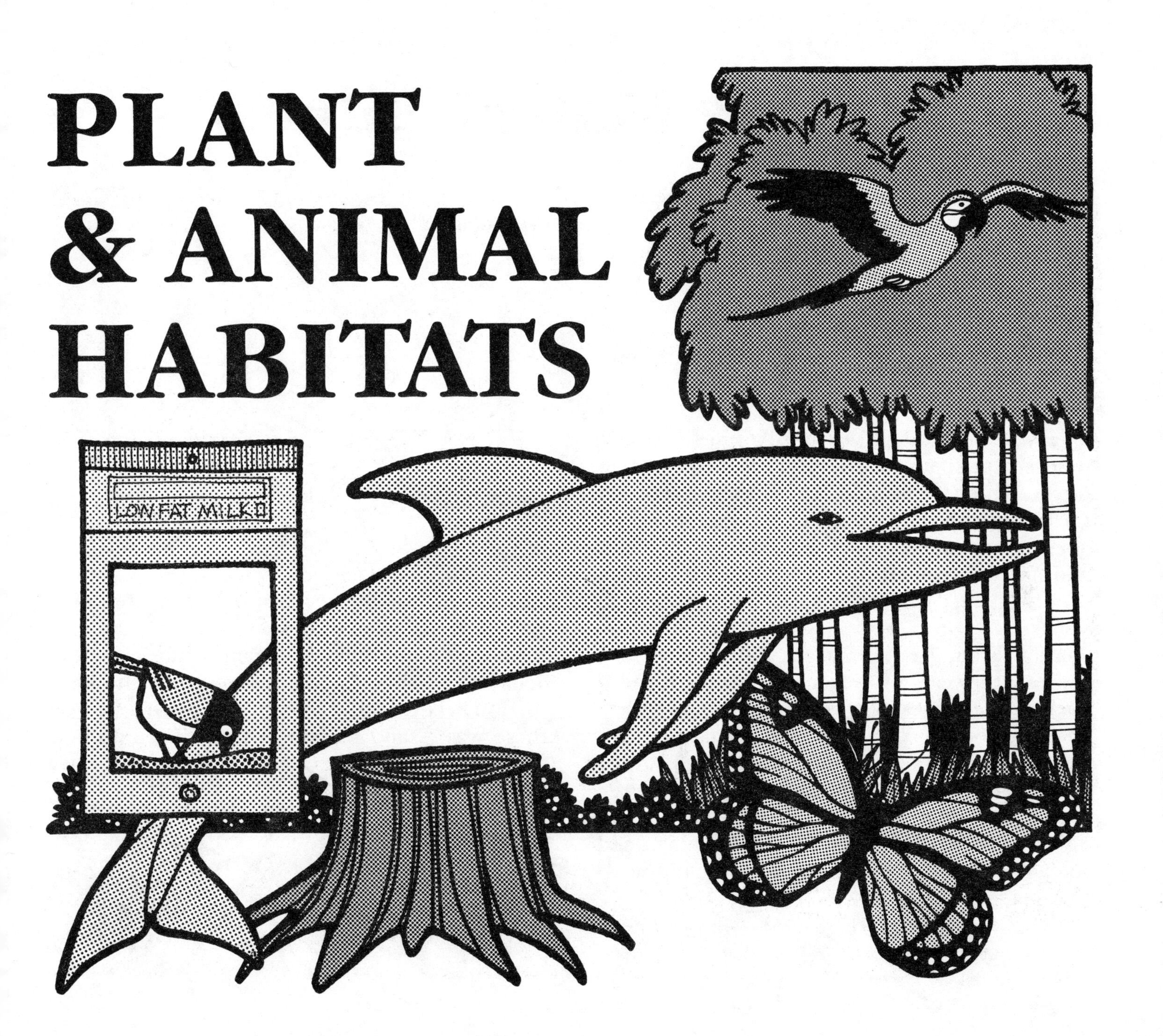
PLANT
& ANIMAL
HABITATS
LOW FAT MILK

The Web of Life

All living things are united in a web of life. Each strand in this web is important. Each plant and animal has a job to do. Each living thing helps maintain the balance of nature and ensure the overall quality of life.

All animals (including people) depend on plants for the air they breathe. When animals breathe out, they exhale a gas called **carbon dioxide**. In large amounts, this gas is poisonous to them. If it built up in the air, it would eventually kill them. But it doesn't build up. Instead, plants take the carbon dioxide from the air and use it to make food. In turn, they give off a gas called **oxygen** which all animals (including people) need. In this way, plants and animals are interdependent. Working together, they maintain a balance that enables them both to survive.

Animals and people also depend on plants for the food they eat. A substance called **chlorophyll** enables the leaves of green plants to use solar energy, carbon dioxide, water, and organic nutrients from the soil to make sugars and starches. Stored in the leaves and stems of the plant, these sugars and starches nourish the plant so it can grow. They also nourish the many animals that eat plants and the wide variety of roots, leaves, fruits, and seeds that plants produce.

As animals digest their food, they produce wastes, parts of the food their bodies do not need or cannot use. If these wastes built up, they would

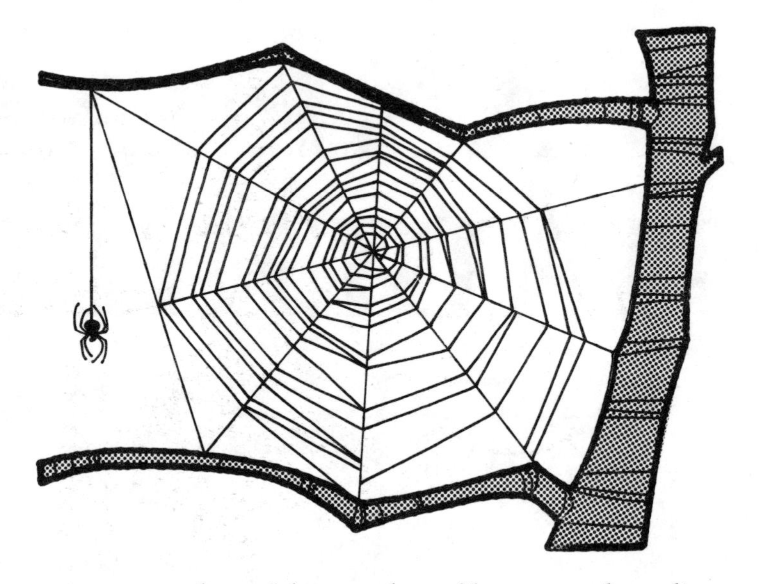

create an unbearable stench and become a breeding ground for all sorts of disease-causing organisms. But they do not build up. Again, nature works to achieve a balance. Decomposers like mushrooms, carrion beetles, other insects, and worms feed on these wastes and transform them into nutrients plants can use. Thus, in nature, nothing is extra and nothing is wasted.

The Web of Life
(continued)

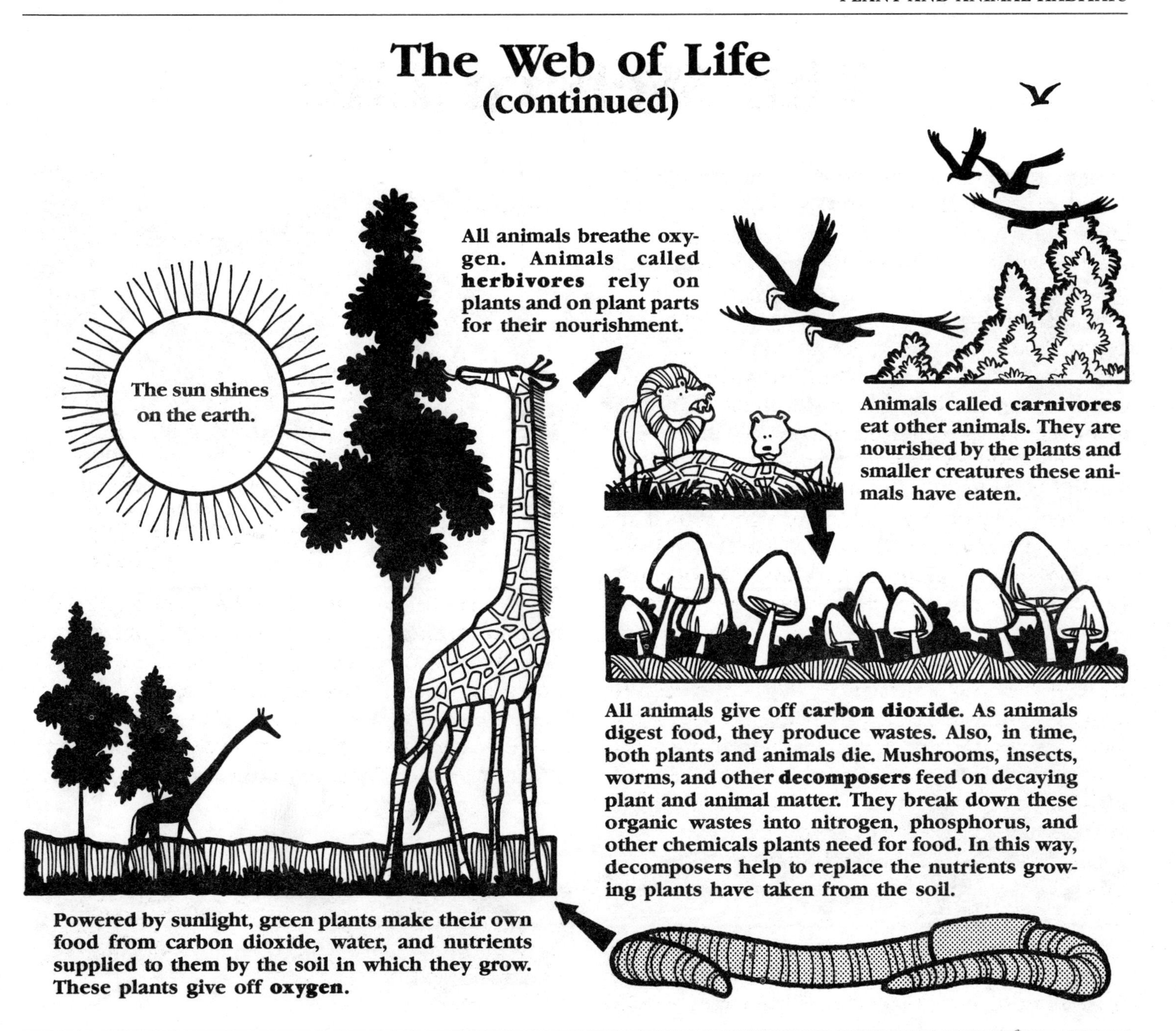

Deforestation

- **Deforestation** is the complete destruction of all forests in an entire region. It is cutting trees with no regard for the land and no effort to replant.
- Deforestation lowers air quality by substantially reducing the number of trees that are available to replenish the oxygen supply.

- Deforestation destroys acres of plant and animal habitats.
- Without habitats, some plants and animals become extinct.

Deforestation
(continued)

- Where trees have been cut, rainwater erodes the exposed soil.
- The nutrient rich **topsoil** is washed away.

- Because shade-giving trees and moisture-holding organic matter have been lost, the air becomes hotter and drier.
- Thus, deforestation may lead to **global warming**, an overall increase in the earth's temperature.

Plant a Tree or Garden

Plant a tree or garden. Select a spot in your yard that is large enough. Be sure that this spot receives adequate sunlight and that water drains from it readily. Condition the soil by loosening it and by adding organic material from your compost (see page 112). To select appropriate trees or plants, read about what they need and how large they become in the information printed on seed packets and plant labels or in a garden book. For best results, follow all instructions carefully.

Tips for Gardeners

- **Weed often.** Weeds take food from your plants and interfere with their growth.
- **Cultivate gently.** Using a hoe or trowel, loosen the soil around the plants. If the soil is packed too tightly, it is hard for food and water to reach plant roots and hard for plant roots to find space in which to grow and spread.
- **Water frequently.** You may need to water soil containing seeds or young plants as often as twice a day to keep it moist. You may need to water soil containing older plants with deeper roots only once a week.
- **Feed regularly.** Growing plants need a balanced diet of nitrogen, phosphorus, and potassium. Read the label to be sure that the food you buy contains what your plants need. Follow the instructions so that you will know how often to feed your plants and how much food to give them.

Fact

The most massive living thing on earth is a tree in California's Sequoia National Park. Named General Sherman, this tree weighs approximately 6,720 tons.

Fact

The leaves of trees and other green plants produce oxygen, the gas people and animals breathe.

Fact

The shade provided by one well-placed tree can reduce the energy needed to air-condition a home 10 to 50 percent.

Enjoying Nature

Take time to enjoy nature. Go to the beach, visit a nearby park, climb to the top of a mountain, walk through a botanical garden, or just sit quietly in your own backyard. Take along a pencil and a sketch pad. Look for things that make you wonder. Notice colors, patterns, shapes, textures, and designs. Draw a picture of something that pleases you, such as a flower, a leaf, a tree branch, or a spider's web. Write a special thought—or even a haiku—about the thing you drew.

Haiku

A **haiku** is an unrhymed Japanese poem. It consists of three lines containing five, seven, and five syllables. A haiku is usually light and delicate in feeling and is most often written about something lovely in nature.

In the sky so blue
Wisps of clouds float gently by
Shading out the sun.

Bird lands on my lawn,
Gathers twigs and blades of grass,
Proudly builds her nest.

EARTHWORDS

Never say there is nothing beautiful in the world any more. There is always something to make you wonder in the shape of a tree, the trembling of a leaf.

—Albert Schweitzer

Build a Compost Heap

Composting is the process of turning organic material you would normally throw away—including grass cuttings, coffee grounds, and grapefruit rinds—into a rich mixture that can be used to condition soil and feed plants. In a compost heap, billions of tiny organisms break down these organic wastes so that they can be used to add nutrients to soil and improve its ability to hold both air and water.

WHAT YOU NEED

- ☐ four wooden corner posts about 4 to 6 feet long
- ☐ enough wooden planks to build side walls 4 to 5 feet high
- ☐ a hammer
- ☐ some nails
- ☐ wooden twigs
- ☐ yard rakings, including cut grass, leaves, and weeds
- ☐ fruit and vegetable kitchen wastes, including apple cores, coffee grounds, carrot and potato peelings, and citrus fruit rinds (Do *not* use beef, chicken, fish, or pork scraps.)
- ☐ kitchen scissors
- ☐ some manure (to speed the decay process)
- ☐ soil
- ☐ a shovel
- ☐ water
- ☐ a sheet of plastic

WHAT YOU DO

1. Choose a place to build your compost heap. This place should measure about 2 square yards.
2. Hammer the four corner posts into the ground.
3. To make side walls, nail the planks to the posts, leaving small amounts of space between for ventilation.
4. Scatter wooden twigs over the ground inside the enclosure.
5. Use kitchen scissors to cut grapefruit rinds and any other pieces of kitchen waste into thin slices.
6. Layer yard rakings, prepared kitchen wastes, manure, and soil over the twigs.
7. Using a shovel, pack the layers firmly.
8. Moisten the layers with water.
9. Lay a sheet of plastic over your compost heap to help trap heat and speed the decay process.
10. Add water regularly to keep the compost heap damp.
11. Once a week, use a shovel to mix the compost material to allow air to circulate and prevent odors from building up.
12. After four to six months, the compost in your heap will be ready for use to condition soil before you plant a tree or garden (see page 110).

For additional information, send a self-addressed, stamped envelope to the Ecology Center, 2530 San Pablo Avenue, Berkeley, California, 94702, and request a Composting Fact Sheet.

What Are Wetlands?

Wetlands are low-lying areas that are saturated with moisture. They can be bogs, deltas, lakes, marshes, or ponds. They provide food-rich habitats for a wide variety of animals, including alligators, bullfrogs, crayfish, muskrats, otters, shellfish, and wading birds.

People once thought that wetlands were useless because they were not suitable for growing crops or building homes. Wetlands were considered to be nothing more than breeding grounds for mosquitoes. These soggy areas were nuisances to be avoided or ignored.

Then people found ways to drain wetlands and fill them. They built shopping centers and parking lots where once cattails had stood and redwing blackbirds had nested. Of the 213 million acres of wetlands that had once existed in the United States, less than half remain.

Today, people are realizing that wetlands are important. They help to purify our water. They prevent flooding and loss of valuable topsoil through erosion. And they are home to nearly one-half of the animals and one-third of the plants on endangered lists.

Marjory Stoneman Douglas

1890–

"There are no other Everglades in the world," announced Marjory Stoneman Douglas proudly in her book entitled *The Everglades: River of Grass.* On the pages that followed, Marjory described a vast wetland on the southern end of the Florida peninsula. She said this sheet of fresh water 6 inches deep and 50 miles wide prevented South Florida from becoming a desert. Marjory's book changed the way people thought about the Everglades. This area, which had once been condemned as a swamp, was declared a national park.

What Are Rain Forests?

Rain forests are tropical woodlands with an annual rainfall of at least 100 inches. These forests grow in the hot, humid areas of the earth, near its equator. In most of these forests, a dense growth of broad-leafed evergreen trees forms an unbroken canopy overhead and provides a shady shelter for the more delicate flowers and vines that grow beneath it.

Rain forests are important for several reasons. They clean and renew the earth's atmosphere by taking carbon dioxide from the air and putting oxygen into it. They are home to many of the earth's animals and plants. And they affect the climate of surrounding areas.

In many countries around the world, people are destroying the rain forests. They want to use the wood from these forests as a building material or fuel. They want to use the land on which these forests grow for homes and factories, ranches and roads. They do not understand that rain forests are essential to the health of the earth and that, once cut, they cannot be regrown or renewed.

Margaret Mee
1909–1989

Margaret Mee was an English artist. She lived in Rio de Janeiro, Brazil. Beginning in 1956, she completed fifteen journeys along the Amazon River. On each journey, she sketched and painted pictures of the flowers she found in the surrounding rain forest. Her sketches and paintings were published in two volumes entitled *Flowers of the Brazilian Forests* and *Flowers of the Amazon*. Margaret's beautiful pictures gave scientists their first glimpse of many remarkable rain forest plants.

What Are Rain Forests?
(continued)

The Amazon rain forests produce 40 percent of the world's oxygen.

Rain forests are home to about one-half of all the plant and animal species on earth.

South American rain forests are the home of the wild cocoa plant, from which chocolate is made.

Rain forest plants are estimated to be the source of ingredients used in 25 percent of the drugs U.S. doctors prescribe.

About 70 percent of the plants used to make drugs for cancer treatment are found only in rain forests.

Throughout the world, rain forests are being destroyed at the alarming rate of 115 square miles a day!

What Is a Habitat?

A **habitat** is the place where a plant or animal naturally lives and grows. For example, a rain forest is the habitat of monkeys. Wetlands are the habitat of frogs. Oceans are the habitat of dolphins and whales. And deserts are the habitat of lizards, snakes, and desert rats. Habitats provide food, water, space to live, and shelter for an interdependent community of living things, which includes both plants and animals.

Different Kinds of Habitats

Oh, Give Me a Habitat!

Pretend that you work for a newspaper called the *Animal Advocate.* It is your job to help animals find suitable places to live. You write real estate ads describing the homes animals want and the specific neighborhoods, or habitats, they require.

First, read the sample ads that appear on this page. Next, select five animals from the Client List. Then, do research to learn more about the needs of these animals. Finally, write an ad for each of these animals in which you describe both the home and the community this animal is seeking.

Client List

Amazon River dolphin
Asian elephant
black-footed ferret
California condor
gorilla
gray wolf
hawksbill sea turtle
Jamaican boa
kiwi
koala
Komodo dragon
kookaburra
oriental white stork
red-necked parrot
right whale
snow leopard

Wanted

A North American backyard for spring and summer occupancy. Prefer site that offers both earthworms and insects. Birdbath a bonus; shade tree a must.

robin

Wanted

A dense mountain forest in China that offers a plentiful supply of bamboo.

giant panda

Wanted

A desert hideaway in which to escape from off-roaders and bikers. Need peace, quiet, and space.

desert tortoise

What Are Endangered Animals?

Over the years, the total population of most kinds of animals remains stable. **Endangered animals** are specific kinds of animals whose total population is becoming steadily smaller, or decreasing. These animals are called "endangered" because they are in danger of dying out completely, or becoming **extinct**.

In 1966, the U.S. Congress officially recognized our responsibility to protect animals that are threatened with extinction. In that year, it passed the Endangered Species Preservation Act. This Act provides for the identification and protection of endangered animals.

Hawksbill turtle

THE TURTLE NEEDS ITS SHELL MORE THAN YOU DO!

Don't buy products made from turtles.

According to the Endangered Species Preservation Act, an animal that has been identified as endangered may not be purchased or sold in interstate or foreign commerce. Also, this animal may not be hunted, shot, pursued, harmed, harassed, trapped, wounded, captured, or collected.

Today, more than 300 national wildlife refuges throughout the United States are safe havens for endangered animals, but the threat of extinction remains. Over 100 names appear on the list of endangered native animals. Some of these names are shown on pages 120 and 121.

Choose an endangered animal from those listed on pages 120 and 121. Do research to learn more about the ways in which the existence of this animal is threatened. For example, you might consider hunting, habitat loss, and pollution. Then design a poster showing a picture of the animal and suggesting at least one specific step people can take to protect it.

EARTHWORDS

The friend of nature is the man who feels himself inwardly united with everything that lives in nature, who shares the fate of all creatures, helps them when he can in their pain and need, and as far as possible avoids injuring or taking life.

—Albert Schweitzer

Why Animals Become Endangered

Animals become endangered for many reasons.

Collection

The **thick-billed parrot** is caught and sold to pet shops.

Hunting

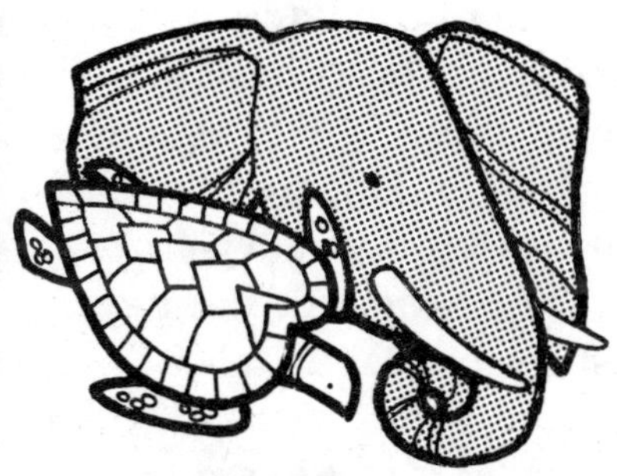

African bull elephants are hunted and killed for the ivory in their tusks. **Hawksbill turtles** are killed for their shells.

Poisoning

Many **California condors** died after eating poisoned meat put out by ranchers to kill coyotes.

Habitat Loss

Jungles that were home to many wild animals have been cut and cleared to provide grazing land for cattle.

Limited Habitat

The **Devil's Hole pupfish** is found only in one small pool in the middle of the Nevada desert.

Introduced Predators

Rats brought by ship to the Galapagos Islands ate the eggs of the **Galapagos tortoise**.

List of Endangered Animals

Below is a partial list of endangered animals. It is based on a list entitled *Endangered and Threatened Wildlife and Plants*, which was compiled by scientists at the Fish and Wildlife Service of the United States Department of the Interior. If you would like to receive a copy of the complete list, you can request one from the

Office of Endangered Species
Fish and Wildlife Service
U.S. Department of the Interior
Washington, D.C. 20240

AMPHIBIANS

Coqui, golden
Salamander, desert slender
Salamander, Texas blind
Toad, Wyoming

BIRDS

Albatross, short-tailed
Bobwhite, masked (quail)
Condor, California
Coot, Hawaiian
Crane, whooping
Eagle, bald
Falcon, peregrine
Goose, Aleutian Canada
Macaw, indigo
Ostrich, West African
Parrot, thick-billed
Pelican, brown
Sparrow, Florida grasshopper
Stork, oriental white

FISHES

Cavefish, Alabama
Darter, Maryland
Killifish, Pahrump
Pupfish, Devil's Hole
Stickleback, unarmored threespine
Sturgeon, shortnose
Trout, Arizona

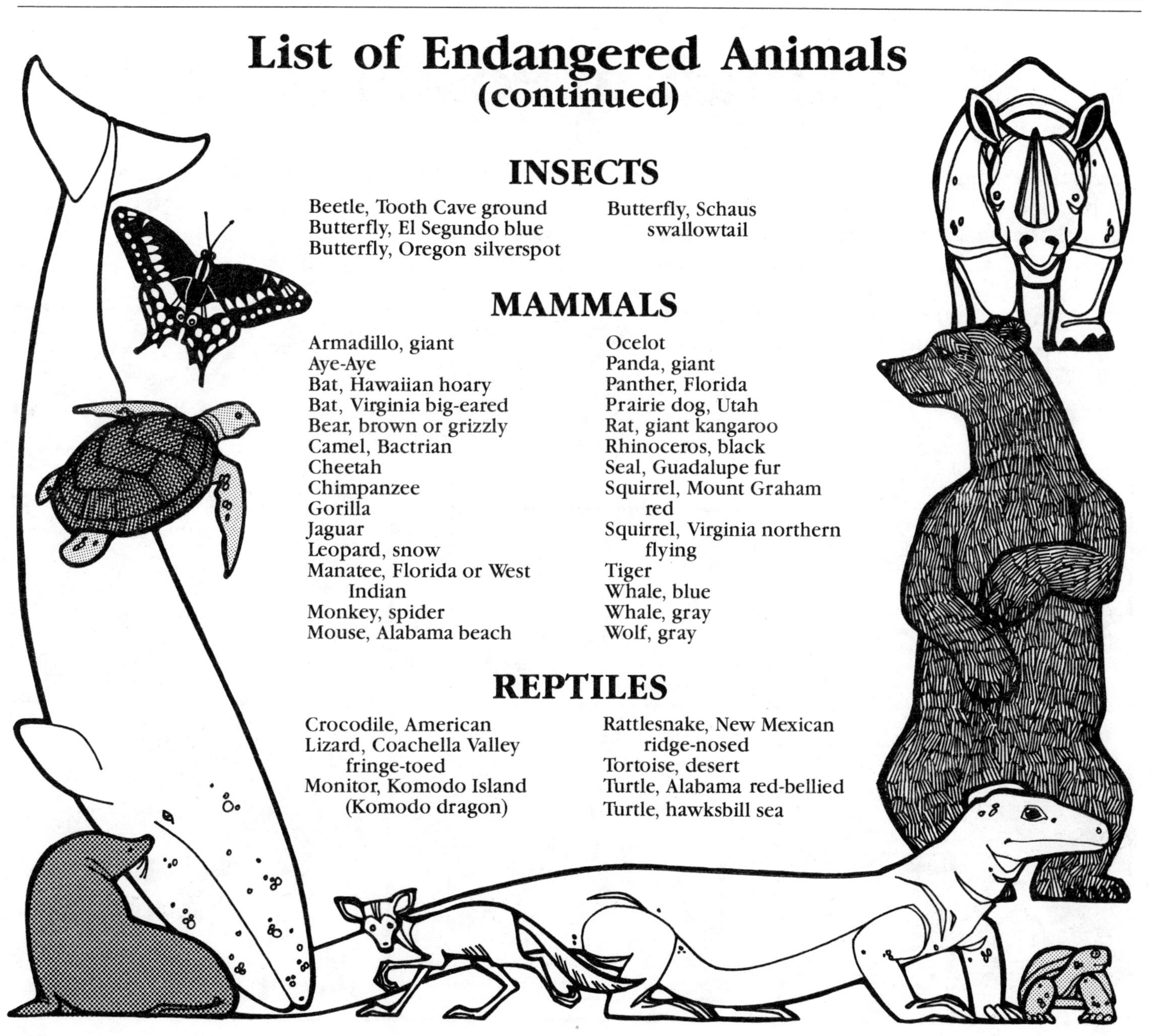

List of Endangered Animals (continued)

INSECTS

Beetle, Tooth Cave ground
Butterfly, El Segundo blue
Butterfly, Oregon silverspot
Butterfly, Schaus swallowtail

MAMMALS

Armadillo, giant
Aye-Aye
Bat, Hawaiian hoary
Bat, Virginia big-eared
Bear, brown or grizzly
Camel, Bactrian
Cheetah
Chimpanzee
Gorilla
Jaguar
Leopard, snow
Manatee, Florida or West Indian
Monkey, spider
Mouse, Alabama beach
Ocelot
Panda, giant
Panther, Florida
Prairie dog, Utah
Rat, giant kangaroo
Rhinoceros, black
Seal, Guadalupe fur
Squirrel, Mount Graham red
Squirrel, Virginia northern flying
Tiger
Whale, blue
Whale, gray
Wolf, gray

REPTILES

Crocodile, American
Lizard, Coachella Valley fringe-toed
Monitor, Komodo Island (Komodo dragon)
Rattlesnake, New Mexican ridge-nosed
Tortoise, desert
Turtle, Alabama red-bellied
Turtle, hawksbill sea

Plan a Poster

Select an endangered animal from the partial list on pages 120–121. Do research in your school or public library to learn more about this animal. Make a poster similar to the one shown below. Using what you have learned, draw or write an answer for each question. Share your poster and your knowledge with others.

How Are Animals Protected?

Natural habitats are protected to ensure that they will provide enough food, water, space to live, and shelter for the animals living within them.

Laws are passed to limit fishing and hunting to certain animals, areas, and seasons. Laws also forbid hunting and killing of animals that are endangered.

Protected areas are set aside in which wild animals and plants can survive. There are more than 1,200 national parks, wildlife preserves, and other protected areas throughout the world.

Shelters are established to house in captivity animals that can no longer survive in the wild. For example, at the San Diego Zoo, California condor eggs are being incubated and hatched in captivity, where the condor chicks will have a much better chance of surviving.

Where the Wild Things Shouldn't Be

How would you like to live in a cage
That was just about ten feet square,
With no toys to play with and nothing to do,
Just you and a bed and a chair?
Oh, sure you'd be fed (the same thing each day),
You'd have water (unless they forgot),
And since you would never be going outside,
You wouldn't get cold or too hot.
But oh, you'd be lonely, just sitting alone
With no one to talk to all day.
You'd remember the trees and the grass and the breeze,
The places where you used to play.
You'd remember your friends, you'd remember the sky,
And games and strawberries and sun,
And you'd know you could never go skating again
Or go swimming or ride bikes or run.
You'd get mad and scream and throw things around;
You'd kick and you'd pound on the wall.
And your owners would scold you and say to themselves,
"He isn't a nice pet at all!"
The more you got mad, the less they would like you,
The less they'd remember to care
About if you had water or if you got fed
Or if you were lonely in there.
And then you would know what it's like to be kept
As a pet when you're meant to be free,
And you'd listen when wild things are trying to say,
"Please don't make a pet out of me."

—Beverly Armstrong

Create an Extinct*less* Animal

Humans contribute to the endangerment and eventual extinction of animals by hunting, by using pesticides, and by destroying their habitats. For example, elephants are hunted for their ivory tusks; other animals are hunted for their skins or shells. Pesticides poison the food or environment of birds, fishes, and other animals that are not considered to be pests. Ranches and housing developments limit the land that is available to support bear, mountain lion, and wolf populations.

Consider the specific characteristics and habitat requirements that make different animals susceptible to human interference. Try to create an animal that would be resistant to the thoughtless actions of humans. On a separate sheet of paper, draw a picture of this imaginary animal and give it a name. Describe its habitat and tell what it likes to eat. List all of the characteristics that protect this creature from the carelessness of humans. As you do so, think about the fact that a characteristic that protects the animal in one way may make it more vulnerable in another way.

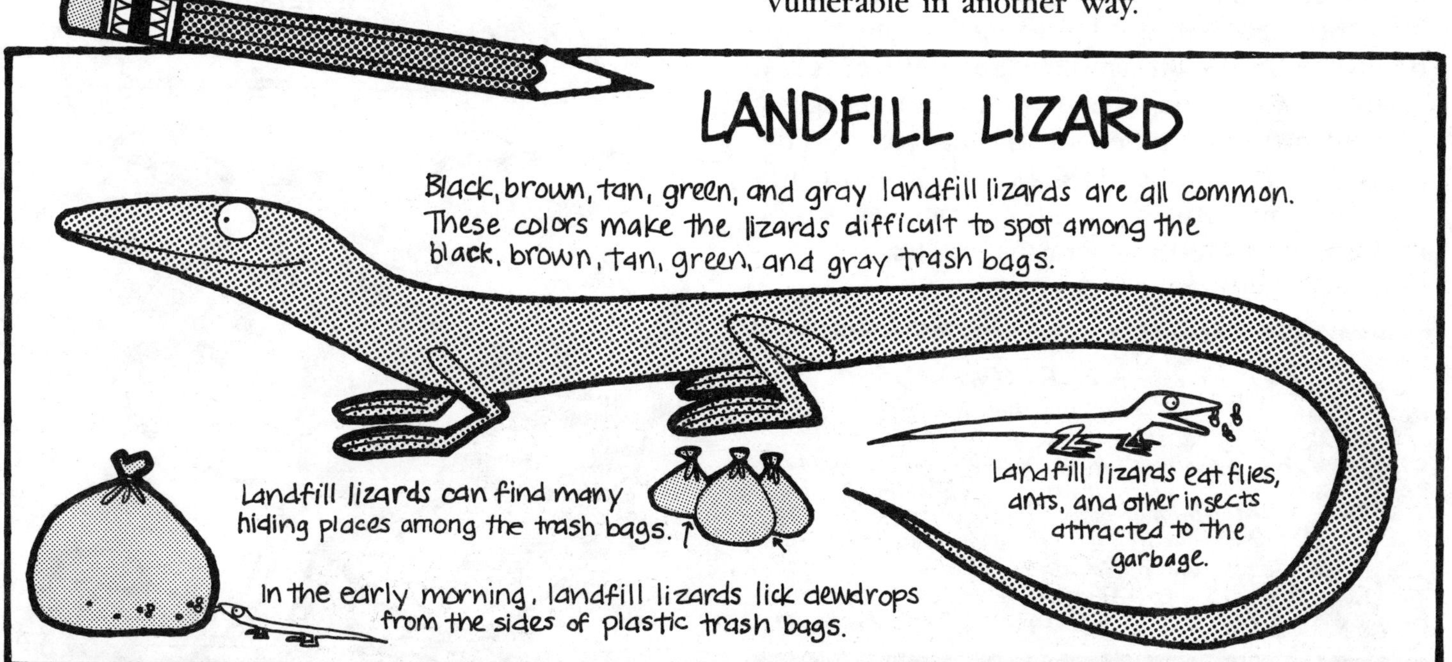

Be a Bird-Watcher

Bird-watching is fun. It gives you a chance to be outdoors and to get close to nature. It also provides you with an opportunity to learn more about how birds look, where they live, what they eat, and how they sound. The best times of day to watch birds are either in the early morning, just before sunrise, or in the early evening, just after sunset.

When you go bird watching, take along a pair of binoculars, a pocket-sized notebook, and a pencil. The binoculars should have a sturdy nylon or leather strap. Place the strap around your neck so the binoculars will not fall to the ground even if they slip out of your hand and so your hands will be free to take notes when you are not using the binoculars. In your notebook, make sketches of the birds you see. Note their sizes, colors, and markings. You may also wish to describe their songs, their flight and feeding patterns, and any other interesting behavior you observe.

EARTHWORDS

The song rules the cloudy dawn, the waiting ranges of hills and their woods full of shadows yet crested with gold, their lawns of light, the soft distended gray clouds all over the sky through which the white sun looks on the world and is glad.

—Edward Thomas
The Song of the Nightingale

Tips

1. **Stand still or move slowly and quietly. Loud noises and rapid movements will frighten away the birds you are trying to watch.**
2. **To observe, find a comfortable position in which you can stand or sit very still for a long time, and be patient.**
3. **While observing, notice the size, shape, color, and songs of the birds you see.**
4. **Buy or borrow a field guide or other book to help you identify the birds you see and learn more about them.**

Milk Carton Bird Feeder

Make a bird feeder from an empty milk carton.

WHAT YOU NEED

- ☐ an empty half-gallon milk carton
- ☐ a pair of sharp scissors
- ☐ a stick 9 to 12 inches long
- ☐ a large paper clip
- ☐ some birdseed

WHAT YOU DO

1. Wash and dry the milk carton.
2. Using the scissors, cut windows in two sides of the carton, leaving margins as shown.
3. Carefully cut a small hole in each of the same sides near the bottom of the feeder.
4. Run the stick through both of these holes to create a perch.
5. Unbend the paper clip to make a hook.
6. Insert this hook through the top of the milk carton as shown.
7. Put some birdseed on the bottom of the feeder.
8. Hang the feeder outside. Choose a spot where feeding birds will be safe from cats and other predators.

EARTHWORDS

There was a strange stillness. . . . The feeding stations in backyards were deserted. . . . It was a spring without voices. On the morning that had once throbbed with the dawn chorus of robins, catbirds, doves, jays, wrens, and scores of other bird voices there was now no sound; only silence lay over the fields and woods and marsh.

—Rachel Carson
Silent Spring

Unusual Pets

Pet stores sell many animals like cats, dogs, and rabbits that have been raised especially to be household pets. Sometimes these stores also sell unusual or exotic animals that have been taken from the wild by trappers. People who buy these exotic animals may be hastening their extinction. For example, some kinds of parrots are endangered. It is illegal to bring these parrots into the United States. Yet greedy animal importers attempt to smuggle these birds across the border and sell them for profit. For every parrot that makes it safely into a family home, many die along the way.

Parrots are caught in nets. As many as two of every five birds caught in these nets die from shock.

Captured parrots are placed in cages for shipping. About one in five dies while caged because of improper feeding and neglect.

To keep the parrots quiet while they are being smuggled, their feet are tied together and their beaks are taped shut. Many die from rough handling or lack of air.

Before you buy an unusual or exotic animal to be a household pet, ask the pet store owner to show you proof that the animal has been raised in captivity. Also do some research to learn about the special needs of this animal. Will you really be able to provide the space, food, and conditions that are necessary to keep this animal healthy and happy? If not, a more ordinary animal might be a much better pet for you and your family.

A NOTE ABOUT COLLECTING

All wild animals have a job to do where they are. They are an important part of the environment in that place. Look at them and learn about them in their natural habitats. Do not capture them and take them home to live a brief and unhappy life and to die an unnatural death in a bottle, box, or jar.

Animal for a Day

It is illegal to dump plastic into the ocean, but merchant ships routinely empty their garbage into our seas. Some of this garbage is plastic. Plastic that is blown off landfills and garbage barges also finds its way into the oceans. Often, sea creatures mistake plastic objects for food. This mistake can be fatal if swallowed plastic lodges in the animal's throat or blocks its digestive tract. In the North Pacific alone, up to 100,000 marine animals and as many as one million seabirds die each year as a result of swallowing plastic or becoming entangled in it.

1. Select one of the following marine animals: dolphin, manatee, pelican, porpoise, seal, swordfish, turtle, or whale.
2. Do research to learn about this animal's food sources and habitat.
3. Pretend that you *are* the marine animal you selected and write a short story about a problem you have with plastic or other nonbiodegradable trash that has been carelessly tossed into the ocean.
4. Illustrate your story.

Trying Times as a Turtle

My name is Timothy Turtle. I am a loggerhead turtle. I love the warmer parts of the Atlantic Ocean where I make my home.

One of my favorite foods is jellyfish. One day, as I was searching for a juicy jellyfish for dinner, I came across something floating on the surface of the water. Thinking it was a jellyfish, I quickly gobbled it down. Much to my dismay, I discovered that I had swallowed a plastic trash bag. What a stomachache I had!

Luckily, I survived. But one of my friends was not so fortunate. A similar plastic trash bag clogged his stomach and killed him. I wish people would be more careful. I don't discard my trash in their homes. Why do they put their trash in mine?

Pick an Animal Project

MAKE A MOBILE

Turn a wire coat hanger into an endangered animal mobile. Choose an animal from those listed on pages 120–121. Cut ten squares from the sides of a brown paper grocery bag. To stiffen, glue pairs of squares together so that the printed sides do *not* show. When the squares are dry, draw or paste pictures of the animal and/or write facts about the animal on both sides of each square. Using thread, hang the squares from the hanger at different heights. Vary their appearance by hanging at least two squares from a corner so that they look like diamonds.

ANIMAL DIORAMA

The museum in your city has hired you to make a diorama showing an endangered animal in its natural habitat. Select an animal from those listed on pages 120–121. Do research to learn more about the habits and habitat of this animal. Then, using what you have learned, turn an old shoe box into a new diorama depicting this animal. For ideas about how to make a shoe box diorama, see page 170.

EARTHWORDS

When we try to pick out anything by itself, we find it hitched to everything else in the universe.

—John Muir

MORE WAYS TO MAKE EVERY DAY EARTH DAY

Thinking About the Earth

Today, the earth faces many problems. It is impossible to read a newspaper or to hear a news broadcast without learning about an environmental problem of some kind. Sometimes, the situation seems hopeless, but you can make a difference if you focus your energies. Start by listing environmental problems that concern you. Here are some general topics that are discussed in this book. Add others you can think of.

acid rain
air pollution
deforestation
endangered animals
litter
water pollution

List specific examples of these problems that affect your daily life at home and at school.

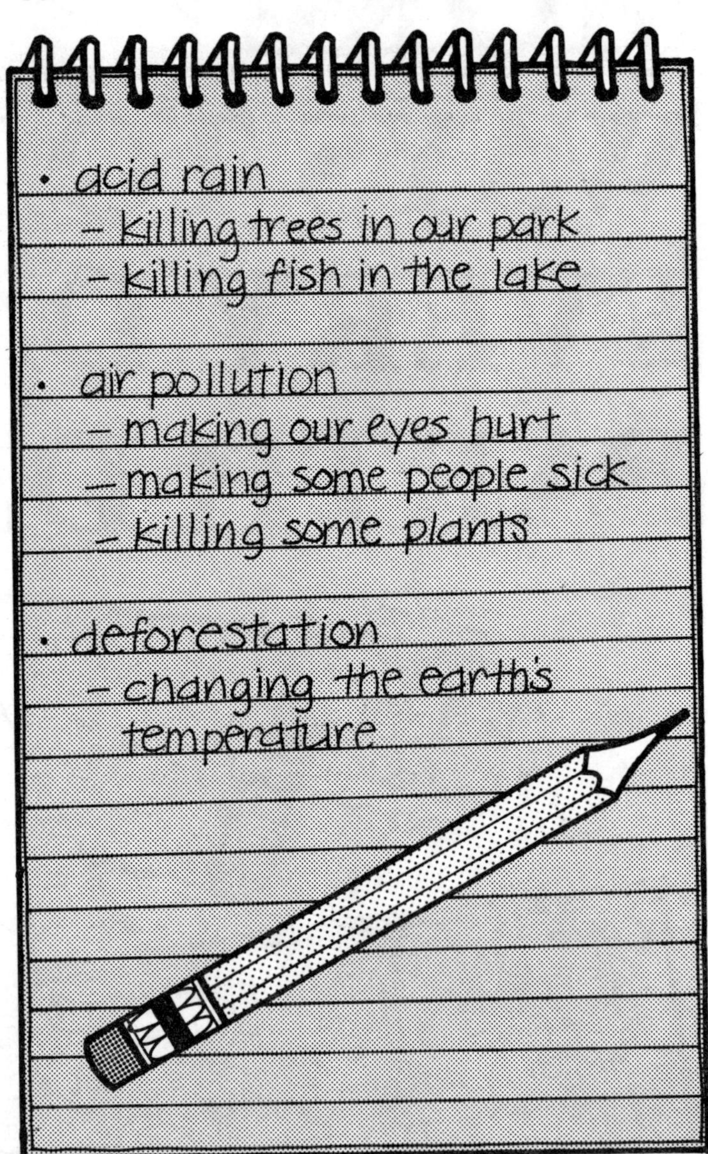

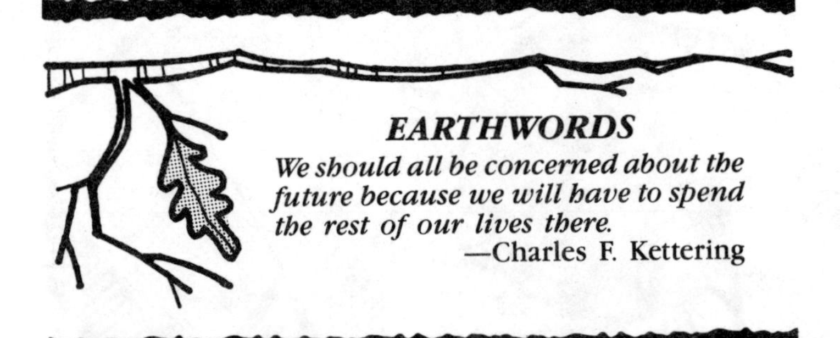

EARTHWORDS

We should all be concerned about the future because we will have to spend the rest of our lives there.
—Charles F. Kettering

Thinking About the Earth
(continued)

Expand your list to include environmental issues that affect your community. Add environmental issues that affect

your state,

your country,

and your world.

EARTHWORDS

A long habit of not thinking a thing wrong gives it the superficial appearance of being right.

—Thomas Paine

Your Community

- trash in Roosevelt Park
- dead fish in Pike's Pond
- loose dogs chasing deer
- landfill almost full
- development in the Back Bay
- improper disposal of motor oil

RESPECT THE EARTH

Your State

- toxic waste near Monarch State Beach
- use of pesticides in agriculture

Your Country

- spilled oil in Alaska
- pollution in the Great Lakes
- endangered Bald Eagles

Your World

- rain forest destruction
- plastic trash in the oceans
- acid rain

Narrowing the Choices

As a part of the activity on pages 132–133, you listed several environmental problems that concern you. For your efforts to be effective, you must focus your attention on one problem at a time instead of worrying about all of them all at once. Here are some questions you might ask yourself when you are narrowing the choices.

Which problem do I think the most about?

Which problem do I know the most about?

Which problem do I care the most about?

In which problem area could I make the biggest difference?

On which problem would I most enjoy working?

Which problem most needs my attention?

Which problem has the greatest effect on my family and me?

Sample Actions

ENDANGERED ANIMALS

As a bird-lover, Michael was most concerned about endangered animals. After writing to some wildlife organizations and reading the information they sent, Michael selected the Western yellow-billed cuckoo as his project. He learned that there may be only fifty breeding pairs of these birds left in the entire state of California. Michael urged kids at his elementary school to write to the U.S. Fish and Wildlife Service asking that this cuckoo be added to the Federal Endangered Species List. Michael stated this same request in a petition, which he placed at a nearby supermarket for adults to sign.

LITTER

Carol was upset that the grounds at her school were always littered with papers, wrappers, and other trash. She persuaded members of her fifth-grade class to take action. First, they created posters to make other kids aware of the problem. Then, under Carol's direction, they formed a litter patrol to give special bookmarks (see page 160) to students who carefully put their litter *in* cans instead of carelessly dropping it *on* the grounds. Finally, Carol and her classmates made certificates (see page 168) to recognize students who picked up ten or more pieces of litter during lunch or recess.

RAIN FORESTS

Steve selected the rain forests as his project because he was concerned that they were being destroyed so rapidly. Steve's goal was to raise money for one of the organizations that is working to save these tropical woodlands. First, Steve created a flier explaining why the rain forests are important and what is happening to them. Then, he organized a bake sale. Steve asked his friends to donate homemade breads, cakes, cookies, and pies. A local merchant agreed to let Steve hold the sale in front of her store. When the sale was over, Steve thanked everyone who had helped and donated all of the money he had raised to the organization he had selected.

Taking Action

Once you have selected an environmental problem, you need to lay out a plan of action. Follow these simple steps.

STEP 1. LEARN ALL YOU CAN ABOUT THE PROBLEM.

Look up the problem category in dictionaries, encyclopedias, and other reference works. Read articles about the specific problem that concerns you in newspapers and magazines. If the problem is a community one, interview experts and/or talk to local authorities. Request related information from some of the organizations listed on pages 173–178. Study the materials they send.

STEP 2. SET A GOAL FOR YOURSELF.

Think about what you hope to accomplish. List three or four specific goals. Be sure that they are workable and realistic.

EARTHWORDS

Even if you're on the right track you'll get run over if you just sit there.
—Will Rogers

Taking Action
(continued)

STEP 3. LIST THE STEPS YOU WILL TAKE.

In making your list, consider the steps suggested in this book. For example, you might

- create posters (see page 169);
- design awards (see pages 85 and 168);
- publish a newsletter (see pages 154–155)
- raise money (see pages 142 and 150);
- volunteer your time; or
- write letters to corporate officers, elected representatives, or the editor of your local newspaper (see pages 156–158).

STEP 4. INVOLVE OTHERS.

Enlist the help of family members, classmates, and friends. By sharing tasks and talents, you can get more done and have more fun doing it.

STEP 5. TAKE THE STEPS YOU HAVE LISTED.

It's time to quit planning and start doing. Commit yourself. Take steps toward your goals.

STEP 6. EVALUATE YOUR RESULTS.

Did your plan work? What have you accomplished? What goals have you achieved? What would you do differently next time?

Celebrating Earth Day

Gaylord Nelson
1916–

Gaylord Nelson, a U.S. senator from Wisconsin, was concerned about conservation. During a speech in Seattle, he proposed a nationwide environmental teach-in on college campuses. Nelson's idea caught on. People in major American cities and in smaller towns joined with students on campuses to hold the first Earth Day on April 22, 1970. In speeches delivered on that day, people learned about the intricate food web that binds all life together and heard Nelson proclaim his goal — "an environment of decency, quality, and mutual respect for all human creatures and for all other living creatures."

". . . the fate of the living planet is the most important issue facing mankind."

—Gaylord Nelson

Earth Day is a special day set aside during the month of April to celebrate our planet. Each year on this day, trees are planted, beaches are cleaned, earth-awareness fairs are held, and people all over the world join together to become more aware of their environment and to help heal the earth.

While it is fun to be a part of these festivities, you should be aware of the environment and do what you can to care for it, not only on one day in April, but on every day throughout the year. The projects and activities in this section will help you stretch the fun and feelings of Earth Day into a lifetime of environmental caring.

Fact

- **In New York City, people celebrating Earth Day on April 22, 1990, left 154.3 tons of litter in Central Park.**

Fact

- **In Baton Rouge, Louisiana, it took city employees and volunteers 3 hours to clean up the trash left behind in 1990 by Earth Day celebrants.**

Making Every Day Earth Day

Write to your state fish and game and/or natural resources department to find out which of the animals that live in your area are endangered. Learn what is being done to protect these animals and how you can help.

Design a crossword puzzle for a friend. Use some of the terms and definitions found in the glossary on pages 179–182.

Invite a guest speaker to talk to your class about the environment.

Start a recycling program at your school.

Volunteer to help an environmental group in your area.

Save energy and natural resources by recycling toys and games. Instead of throwing away the ones you have outgrown or are tired of playing with, organize an exchange. Invite neighbors or classmates to come and trade their used toys and games for ones they've never owned or played with.

Many trees must be cut down to make the paper on which books are printed. You can save some of those trees by establishing a paperback book lending library in your classroom. With your teacher's permission, encourage classmates to bring in paperback books they have read and enjoyed but no longer want. Display these books on a table or shelf, and let students borrow them for in-class leisure reading or to take home on weekends.

Keep your pets at home. Don't let dogs or cats chase or injure birds and other wild animals that may live in or near your community.

When you are enjoying nature, remember the three l's—*look*, *learn*, and *leave it alone*. Don't collect eggs, disturb nests, or remove creatures of any kind from their natural habitats.

Share what you learn about the environment with others. Making every day earth day means caring for the environment in every way you can in all the places you are on every day you live.

Creative Writing Ideas

ENVIRONMENTAL ALPHABET

Print a large letter **A**. Beside this letter, write a word that starts with the letter *A* and is related to the environment. Add a fact or idea about this word. Do this same thing for all twenty-six letters of the alphabet.

A is for **aluminum**. Every year Americans throw away about 35 billion aluminum cans.

B is for **biodegradable**. Biodegradable objects break down readily and can be recycled by nature.

C is for **conservation**, the act of protecting and saving our natural resources.

D is for **disposable diapers**. Each year, Americans throw away about 18 billion disposable diapers. Most of them end up in landfills.

E is for **endangered animals**. Currently, about 235 native plants and animals are listed as endangered.

WORD PICTURE

Go outside and look for something special. It could be a bird building a nest, a butterfly perched on a flower petal, or the gnarled trunk of an old tree. Create a word picture of it. Use poetry or prose, but choose your words carefully so that someone who reads them will be able to see the picture you are trying to create.

WISH FOR THE WORLD

Write a letter to someone who lived fifty years ago describing what you wish he or she had done differently so that the environment would be in better shape today. Mention some of the mistakes that were made. Tell how these mistakes might have been avoided *then*. Also tell what you will do *now* to correct these mistakes and help heal the earth.

THANK-YOU NOTE

Choose a creature whose actions benefit you or the environment in some way. Write a note to this creature in which you describe what it does and thank it for helping. For example, you might write a note to a honey bee thanking it for pollinating fruit tree blossoms so that you will have crisp, juicy apples to eat.

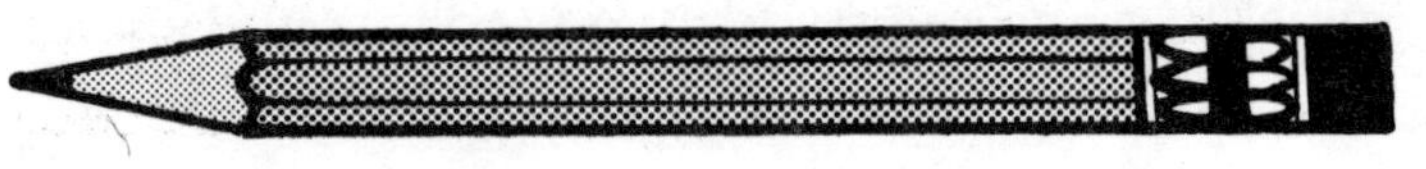

Creative Writing Ideas (continued)

A JINGLE OR LIMERICK

Write a jingle, limerick, or other humorous poem reminding kids of something they can do to help heal the earth. For example, your poem might caution readers not to waste water.

There was a young lad named Lane
Who didn't have much of a brain.
He left tap water rushing
While he did his brushing—
Wasted water went right down the drain.

A GREETING CARD

Create a greeting card to send on a special earth occasion. For example, your card might be in honor of Arbor Day, Be Kind to Animals Week, Cleaner Air Week, Earth Day, or Groundhog Day. Fold a sheet of paper in half. Draw a picture on the front. Write a verse or message inside. Deliver your card in person or put it in an envelope and mail it to a family member, classmate, or friend.

A SERIOUS POEM

Inspire people to help heal the earth by writing a poem in which you describe this planet as something beautiful and worth caring for.

Awakening Planet

We are living
on an
awakening planet
coming to realize that
all the foolish
things
we do
are worthless.
Happiness and
peace
are coming to reign
in a world of snowflakes
and sunshine of cat's eyes
and turtle doves
In a world of life and death
or
both merged together
In a world beyond reality
where dreams are singing
to the wind
In a world
where violets and diamonds
are one and the same
on this
awakening
planet

—Chloe Coventry, Grade 6

Fund-Raising Fun

Alone or with a group of friends, raise money to donate to an environmental organization.

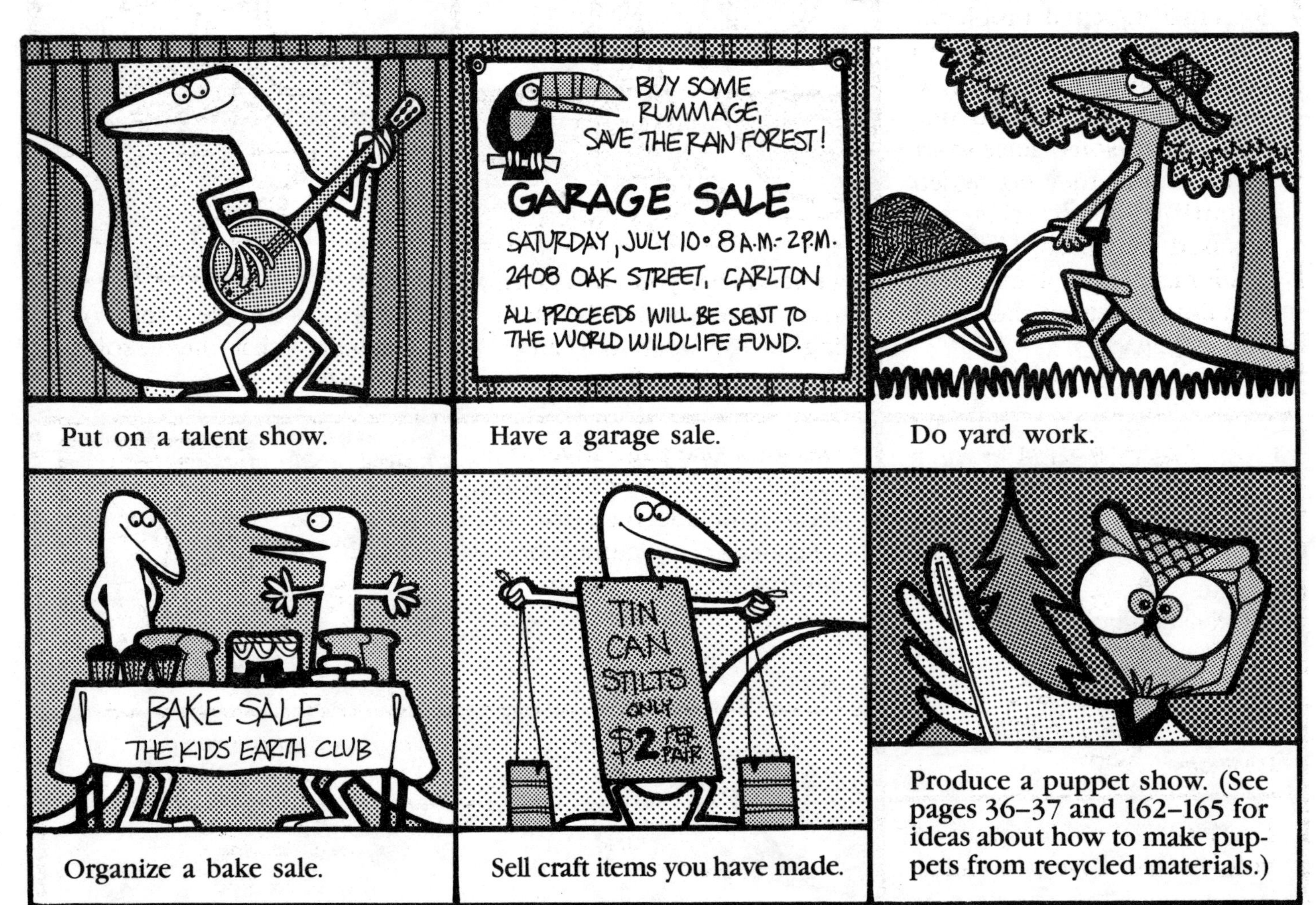

Put on a talent show.

Have a garage sale.

Do yard work.

Organize a bake sale.

Sell craft items you have made.

Produce a puppet show. (See pages 36–37 and 162–165 for ideas about how to make puppets from recycled materials.)

Design a Device

Science is sometimes blamed for environmental problems, but scientists don't set out to create problems. They try to solve problems and to discover the reasons things work or happen as they do. When scientific discoveries are misused or misapplied, the result may be problems scientists did not intend and could not anticipate.

Pretend that you are a scientist or inventor who is searching for the solution to an environmental problem.

Design a device—real or imaginary—that might solve the problem.

Draw and label a diagram of your device. Give it a name.

Design a one-page advertisement for your device. This ad should include the name of your device, a picture, a brief description of how the device works, a list of special features, and the purchase price.

EARTHWORDS

The human future depends on our ability to combine the knowledge of science with the wisdom of wildness.

—Charles Lindbergh

Kids' Views

What is the most serious environmental problem we face today? Take a survey to find out how your classmates and friends would answer this question.

WHAT YOU NEED

- ☐ ten survey forms (one for each person you plan to interview)
- ☐ a pencil or pen to record responses

WHAT YOU DO

1. Interview ten classmates and friends. Ask them to rank the environmental issues listed on the survey form in the order of their seriousness. Have them place a number 5 by the most serious problem, a number 4 by the next most serious problem, and so on.

Sample Survey Form

Instructions: Rank each environmental problem listed below in the order of its seriousness. Place a number 5 on the line in front of the problem you consider to be the most serious, a number 4 on the line in front of the problem you consider to be the next most serious, and so on. Place a number 1 on the line in front of the problem you consider to be the least serious.

_____ air pollution
_____ deforestation
_____ endangered animals
_____ litter
_____ water pollution

You need not sign your name.

2. To analyze the results of your survey, first add up the scores received by each environmental problem. Then, list all five problems in order of their scores, putting the problem with the highest score at the top of your list. This is the problem that kids responding to your survey consider to be the most serious.

Survey Results

Endangered animals	50
Litter	45
Air pollution	25
Water pollution	18
Deforestation	12
Total	150

3. Make a bar graph showing the results of your survey.

Relative Seriousness of Environmental Problems

Problem — **Seriousness Score** 0 10 20 30 40 50

air pollution
deforestation
endangered animals
litter
water pollution

4. Share your results with classmates and friends.

Environmental Scrapbook

Keep an environmental scrapbook. Fill it with interesting magazine and newspaper articles, pictures, interviews, facts, figures, charts, and graphs about the environment. Be sure to include the good news about things people and companies are doing to help heal the earth—not just the bad news about problems—and some of your favorite earthwords.

Create a Board Game

Work alone or with a group of friends to create a board game that will help others become more aware of the environment.

WHAT YOU NEED

- ☐ plenty of scratch paper for your rough drafts
- ☐ several sheets of plain white paper for the game rules and answer sheet
- ☐ a pencil
- ☐ a copy of the Learning Works *Earth Book for Kids* plus dictionaries, encyclopedias, and environmental reference materials from the sources listed on pages 173–178
- ☐ 50 or so 3-inch-by-5-inch index cards
- ☐ a sheet of white tagboard or poster board
- ☐ a ruler
- ☐ crayons or felt-tipped marking pens
- ☐ a pair of scissors
- ☐ an assortment of objects, such as brads, paper clips, pennies, small erasers, toy cars, and the like, to be used as playing pieces
- ☐ a reusable box to hold game cards and playing pieces
- ☐ tape or glue
- ☐ one die or spinner

WHAT YOU DO

1. Decide on a name for your game.
2. On a sheet of scratch paper, draw a rough sketch of the game board.
3. On another sheet of scratch paper, write 30 questions about the environment. Base these questions on facts found in the Learning Works *Earth Book for Kids* and in dictionaries, encyclopedias, and some of the environmental reference materials you have received from the sources listed on pages 173–178. As you write these questions, number each one.

Create a Board Game
(continued)

4. To make an answer sheet, list the question numbers on a separate sheet of paper and write the correct answer beside each one.
5. Reread your questions. When you are satisfied with the way in which they are worded, carefully print each question on a separate index card. Remember to include the question numbers.
6. To make your game more interesting, add some chance cards. For example, one chance card might read: *Ocean polluted. Lose a turn.* Another card might read: *Landfill full. Take trash back 3 spaces.*
7. Make a final draft of your game board on tagboard or poster board. Use a ruler and work carefully. Print the name of your game in large, neat letters. Add pictures to make your game more attractive. Color your board.
8. Find or make playing pieces for your game. Among the ready-made objects you might use are brads, paper clips, pennies, small erasers, and toy cars. Playing pieces should be different so that they can be easily distinguished from one another.
9. Decorate a reusable box to hold game cards and playing pieces. (Your game board need not fit inside this box.)
10. On a sheet of scratch paper, write out the game rules. Be sure that your rules state the answers for these questions.
   a. What is the object of the game?
   b. How many players can play?
   c. Which player goes first?
   d. What does that player do—roll the die, spin the spinner, take a card, answer a question?
   e. How is a player rewarded when he or she answers a question correctly?
   f. How is a player penalized when he or she answers a question incorrectly?
   g. Who checks the answer sheet?
   h. When is the game over?
   i. How is the winner determined?
11. Print a final copy of the game rules.
12. If possible, tape or glue the game rules and the answer sheet to the inside of the box lid so that they will not get lost.

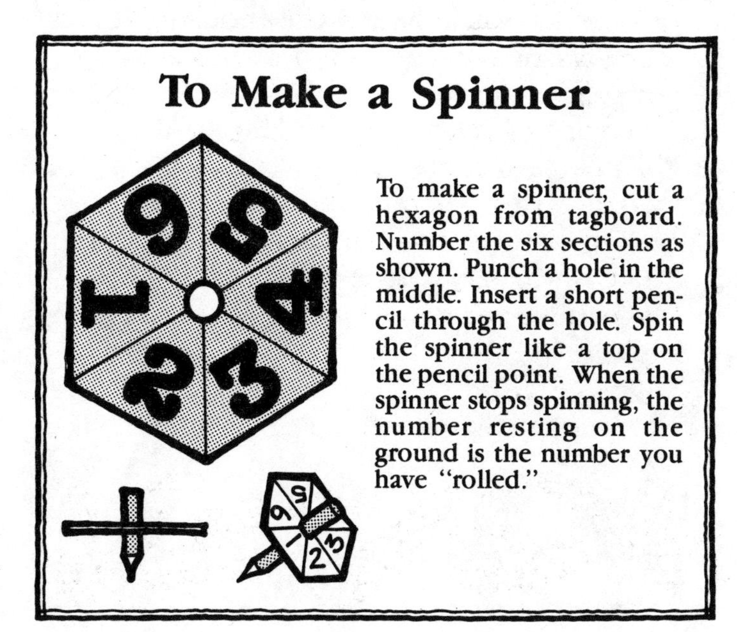

To Make a Spinner

To make a spinner, cut a hexagon from tagboard. Number the six sections as shown. Punch a hole in the middle. Insert a short pencil through the hole. Spin the spinner like a top on the pencil point. When the spinner stops spinning, the number resting on the ground is the number you have "rolled."

Make Your Own Paper

To understand clearly how paper is recycled and reused, make some of your own.

WHAT YOU NEED

- ☐ some newspapers
- ☐ a bucket
- ☐ some water
- ☐ a wire whisk
- ☐ 3 tablespoons of cornstarch
- ☐ measuring spoons
- ☐ a piece of screen that measures about 6 inches across
- ☐ a rolling pin
- ☐ a sheet of plastic wrap large enough to cover the screen

WHAT YOU DO

1. Tear some of the newspapers into small pieces.
2. Put the torn paper pieces into the bucket until it is half full.
3. Add enough water to wet the paper pieces thoroughly.
4. Let the paper-and-water mixture stand for several hours.
5. Using the whisk, beat the mixture into a creamy pulp.
6. Dissolve 3 tablespoons of cornstarch in one cup of water.
7. Add the dissolved cornstarch to the pulp and stir to mix thoroughly.
8. Submerge the piece of screen in the pulp and pull it out.

Make Your Own Paper
(continued)

9. Repeat step 8 until the screen is covered with about a ⅛-inch layer of paper pulp.
10. Spread out some sheets of newspaper.
11. Lay the pulp-covered screen on this newspaper.
12. Cover the screen with a sheet of plastic wrap.
13. Use a rolling pin to press out the excess moisture.
14. Prop the pulp-covered screen up so that air can circulate through it and it will dry.
15. When the pulp is dry, gently peel this sheet of recycled paper from the screen.

WRITING WORDS

Look up each of these words to learn what it means and how it is commonly used.

calligraphy
cursive
hieroglyphics
illumination
inscribe
legible
papyrus
pulp
quill
sans serif
scroll
stylus

WHO?

Who first made and used paper?

WHAT?

What did people write on *before* they had paper?

Pennies for Programs

Instead of recycling one large glass jar, reuse it as a penny bank. Ask family members and friends to empty their pockets and drop their spare pennies into your jar. When your jar gets full, go to the bank and ask a teller for some penny wrappers. These paper cylinders are designed to hold fifty pennies each. Carefully count your pennies and put exactly the right number in each wrapper. Take your wrapped pennies back to the bank and ask the teller to give you one dollar for every two rolls of pennies you have wrapped. Use the money you collect in this way to help the environment. For example, you might use it to plant a tree, to buy advertising, or to print fliers about an environmental cause in which you believe. Consider using it to send letters to members of Congress regarding an environmental or wildlife bill under consideration, or you might donate it to a conservation or wildlife group. For the names of some groups that could use your help, see pages 174–178.

EARTHWORDS

There is never an instant's truce between virtue and vice. Goodness is the only investment that never fails.

—Henry David Thoreau
Walden

Photo Contest

Organize a photo contest in your classroom or community. Not only will this project get the entrants involved, but the displayed photographs may make other people in your school or community more aware of environmental issues.

1. Decide on a theme for the contest. Here are some possible choices.
   - Conservation—Your Concern
   - Energy-Saving Ideas
   - Making Every Day Earth Day
   - Plant and Animal Habitats
   - Pollution Solutions
   - Recycling Round-up

2. Establish some rules and guidelines for the contest. Here are some questions to consider.
   - Who may enter the contest?
   - Will there be different age categories for contest entrants?
   - Will you accept both color and black-and-white photographs?
   - What size must the photographs be?
   - What information should appear on contest entry forms?
   - When does the contest open?
   - When does the contest close? What is the deadline for submitting photographs?
   - Where and to whom should the photo entries be submitted?
   - Who will judge the contest?
   - What criteria will the judges use?
   - What will the prizes be?

3. Attach a label to the back of each submitted photo. This label should give the name of the photographer, the age (if there are different age categories), a brief description of the place photographed or the circumstances under which the photo was taken, and a title.

4. Display the photographs during a school open house or in a public library or community hall.

5. Create posters and fliers to publicize details of the contest and to invite people to view the display.

Take a Tour

With your family, take an environmental awareness tour. Here are some of the places you might visit.

Environmental Issue Poster

1. Choose an environmental issue that is of special interest to you. For example, you might consider air or water pollution, deforestation, endangered animals, energy shortages, diminishing landfill capacity, ozone depletion, water conservation and/or wilderness preservation.
2. From newspapers and magazines, cut out and collect headlines, articles, editorials, and pictures about this issue.
3. Select the best items from your collection and glue them to a piece of cardboard or poster board to create an environmental issue poster.
4. Share your poster with others by displaying it at home or at school.

Publish a Newsletter

One way to make people aware of your environmental concerns is to publish a newsletter about them. You can prepare the original in several ways. For example, you can hand letter it using black ink. You can cut and paste words and pictures from other sources. You can type it on a typewriter, or you can compose it on a computer. Then you can have your finished newsletter photocopied so that you will be able to give or sell copies to others.

1. Think of a name for your newsletter. This name should be no longer than two or three words and should tell people what the newsletter is about. You can use one of the following names or create your own.

Earth Chronicle
The Earth Times
Global Concern
The Planet Protector
Wildlife Watch
The Weekly Wave

2. Display this name in the **nameplate** at the top of the front page of your newsletter. Below the name, print a line showing the volume number, date, and issue number of your newsletter. If you plan to sell copies, you may also want to include the purchase price.

Monthly Planet

VOL. 1 | April 22, 1991 | NO. 1

Protests Persuade Tuna Canners To Protect Dolphins

Protests by concerned kids have persuaded tuna canners to protect dolphins by buying tuna only from fishermen whose fishing methods do not harm these mammals.

For many years, tuna fishermen have been using gill nets to catch tuna. These nets catch lots of tuna, but they also catch dolphins that swim with the tuna.

Dolphins are mammals. They cannot breathe under water. They need air.

When dolphins get caught in gill nets, they cannot dive for food. They cannot swim to the top of the water for air. Without food or air, the dolphins die.

Kids heard about the dolphins dying. They became concerned. They would not eat tuna. They wrote to canning company presidents to tell them why they were not eating tuna.

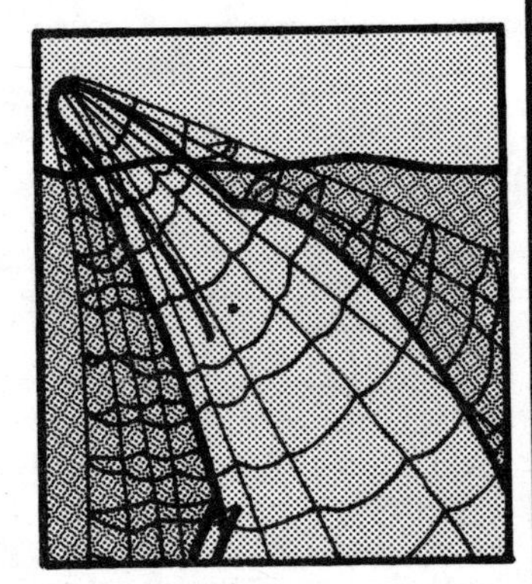

The company presidents read the letters. They cared about dolphins. They decided to buy tuna only from fishermen whose methods do not hurt dolphins.

BUY TUNA FROM CANNERS WHO CARE

Publish a Newsletter
(continued)

Monthly Planet

Publisher	*Linda Schwartz*
Editor in Chief	*Sherri Butterfield*
Illustrator	*Beverly Armstrong*
Layout Specialist	*Margy Brown*
Staff Writers	*Mike Schwartz*
	Stephen Schwartz
Proofreaders	*Bobbe Dartanner*
	Christina Lange

Letters to the Editor

Dear Editor:
My family and I often picnic in Lincoln Park. Lately I've noticed that this park is very dirty. Frankly, it's not because people are careless. It's because there aren't enough trash cans. On busy weekends, the cans get so full that the lids won't close. The wind blows paper napkins and plates out of the trash cans and around in the park. Who runs this park and how can we get them to put more trash cans in it?
Concerned Reader

Dear Reader:
Thank you for your concern. Lincoln Park is owned and operated by the city. I have written letters to various members of the city council asking them to take a look at the trash can problem and see what can be done. I'll let you know what they decide.
Editor

3. Publishing a newsletter can be a big job, so you may want some help. Ask friends and classmates to become members of the **newsletter staff**. Assign duties and list names and titles on your **masthead**.

4. Discuss the purpose of the newsletter with the members of your staff. Decide what topics you will cover and how you will treat these topics. Will you give the facts in a **news story**, give opinions in an **editorial**, or will you write a **feature story** that is designed to inform and entertain your readers?

5. Talk about your **editorial policy**. Establish **guidelines** and set **deadlines**.

6. To make your newsletter more interesting, plan to include special features like drawings, cartoons, crossword puzzles, riddles, games, little-known facts, an ask-the-experts column, and some letters to the editor.

Write a Letter

Writing a letter is a good way to communicate your concerns about the environment to other people. Keep these points in mind when you write.

1. Make your letter simple and direct.
2. Focus on one problem.
3. Organize what you want to say. First, explain why you are writing and what your concern is. Then, express your opinion and offer some suggestions about how the problem might be solved.
4. Type or use your best handwriting so your letter will be neat and easy to read.
5. Before mailing your letter, check it for spelling, punctuation, or other errors, and correct any you find.
6. Sign your name and include your address so that the person to whom you are writing can respond.

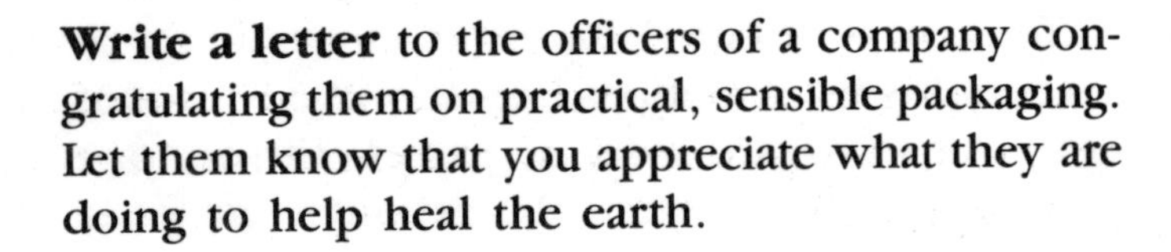

Write a letter to the officers of a company congratulating them on practical, sensible packaging. Let them know that you appreciate what they are doing to help heal the earth.

Write a letter to officials of your local, state, or national government explaining your concerns about the environment. In your letter, ask them what they are currently doing about a specific problem. Find out what legislation is pending on this issue. Share any information you receive with others.

Write a letter to the editor of your local newspaper expressing your feelings and opinions about air pollution, deforestation, endangered animals, oil spills, water pollution or some other environmental problem. In your letter, describe what steps community leaders might take toward solving this problem.

Write a letter to the officers of a company that is not doing its share to help heal the earth. Let them know how you feel. Challenge them to make their manufacturing or packaging processes more earth-friendly.

Write a Letter
(continued)

LETTER-WRITING TIPS

This is the correct form for writing and addressing a business letter.

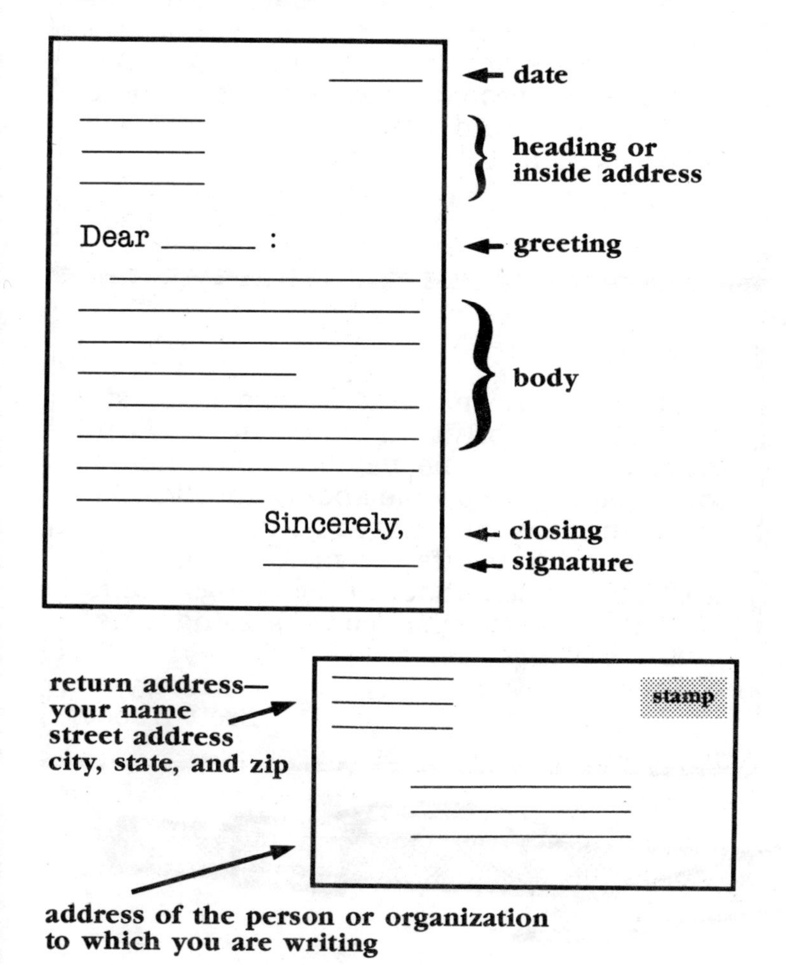

When writing to the president of the United States, address your letter and envelope as follows:

President (name)
The White House
1600 Pennsylvania Avenue
Washington, D.C. 20500

When writing to a United States senator, address your letter and envelope as follows:

The Honorable (senator's name)
U.S. Senate
Washington, D.C. 20510

When writing to a member of the U.S. House of Representatives, address your letter and envelope as follows:

The Honorable (member's name)
U.S. House of Representatives
Washington, D.C. 20515

If you do not know who your United States senators and representatives are, you can find their names listed at your local library.

Kids' Letters

Dear President Bush:

I am very concerned about air pollution in America. I do gymnastics, and my team travels to Los Angeles for meets. I can barely compete down there. My nose runs and my eyes water. I can't imagine what the smog is like in more heavily populated cities like New York. Could you please appeal to Congress concerning new factory and toxic waste laws?

Sincerely,
Julie Dobie

To Whom It May Concern:

My father and I like to go fishing. . . . Recently he caught a fish with plastic bits in its stomach. We were disgusted! Another time my father and I were fishing and a dolphin was stuck in a drift net. My father cut it out and probably saved its life. I would like it if you could write and present a bill to ban drift nets.

Sincerely,
Sarah Kitson

Dear Congressman Lagomarsino:

My concern is the ozone layer. My suggestion is that we limit how many aerosol and gasoline products we buy, . . . or we could get scientists to invent a substitute for CFCs to put in the sprayers.

Sincerely,
Steven West

Dear Senator Hart:

One of my major concerns is plastic wastes and pollution in California and everywhere. . . . One way you could help is to educate more people about our situation and help pass laws about pollution.

One ad that affected me . . . said, "We were so worried about other beings killing us that we forgot what we were doing to us."

Sincerely,
Brett Abbott

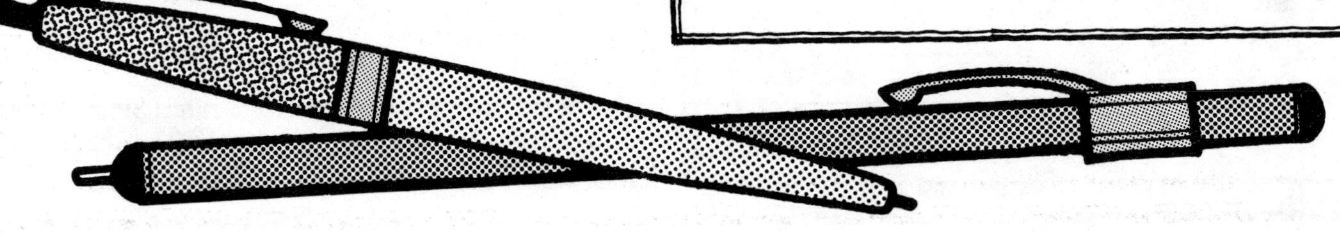

Cartoon Time

Words are one way to express ideas; pictures are another. Use both words and pictures to create a single cartoon or a cartoon strip that expresses some of your ideas about the environment. For example, you might draw a tall-tale-type hero who is able to clean up ocean oil spills with a single, mighty wipe. Or you might show a group of gentle forest creatures discussing pollution problems from *their* point of view. Make your cartoon either serious or funny, but have some fun with it.

Bookmark, Stamp, or T-Shirt

Design a **postage stamp** in honor of the earth. In your design, include a picture of some kind, the value of the stamp, and the name of the country issuing it.

Write some earthwords you would like to wear on the front of a **T-shirt**. Include a symbol or picture.

Using recycled paper, create a **bookmark** encouraging others to join you in caring for the earth.

Banner or Bumper Sticker

BANNER OR FLAG

You have been chosen to create a banner or flag to tell people how you feel about the environment. Your theme can be cleaning up litter, conserving natural resources, protecting endangered animals, saving energy, stopping deforestation, or anything else you care about.

Draw your design on a white piece of paper. Use crayons or felt-tipped marking pens to color it. Select colors that will go with nature and convey your message. Make your banner from felt and hang it from a dowel. Or make a flag from an old bed-sheet and use masking tape to attach it to a broomstick or other pole.

BUMPER STICKER

Bumper stickers must be read quickly, so the messages on them are short and to the point. Think of an environmental message that could appear on a bumper sticker. Carefully print your message on a rectangular piece of poster board or cardboard. Make your letters large and keep your message short. If you have room, add a border, a small picture, or some other decoration.

Recycled Cup Puppets

Recycle used paper or polystyrene foam cups into puppets. Make your puppets the central characters in a play about caring for wildlife and healing the environment.

YOU WILL NEED

- ☐ used paper or styrofoam cups, rinsed and dried
- ☐ a pencil
- ☐ scissors
- ☐ glue
- ☐ cellophane and masking tape
- ☐ a stapler and staples
- ☐ crayons or felt-tipped marking pens
- ☐ recycled odds and ends, such as bits of construction paper, brads, cotton balls, fabric scraps, feathers, felt, paper clips, pieces of yarn, pipe cleaners, and toothpicks

YOU CAN MAKE

BIRDS OR BATS

Make a bird or a bat by drawing a face on your cup and adding wings. If you are making a bird, you may want to add a feather or two. Puppets of this kind can be taped to sticks or hung from pieces of string or elastic.

A RABBIT

Draw a rabbit's face on a paper or polystyrene foam cup. Add cloth or paper ears. For the body, dress your arm in a white sock. Glue felt or paper legs and a cotton puff tail to the sock. Put the cup on your hand to make the rabbit's head.

Recycled Cup Puppets
(continued)

YOU CAN MAKE

TALKING PUPPETS

To make a talking puppet, find two cups with handles. Stack the cups so that the tops are together and the handles are one below the other.

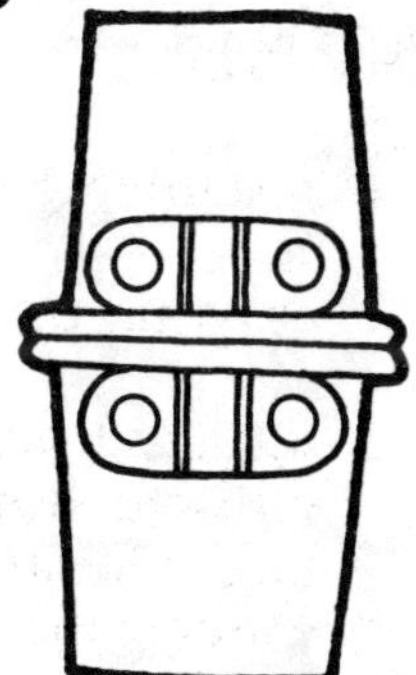

Fasten the cups together by running a strip of masking tape from one cup to the other between the open handles.

Use yarn, paper, and/or marking pens to add hair and facial features to the top cup. Then, fold the handles back and hold your puppet as shown.

STACKS OF ACROBATS

Glue or tape at least five cups together, top to top and bottom to bottom, to make a stack of acrobats. Add arms and legs made from paper or pipe cleaners. Use marking pens to draw hair and facial features, or cut them from paper or yarn and glue them on.

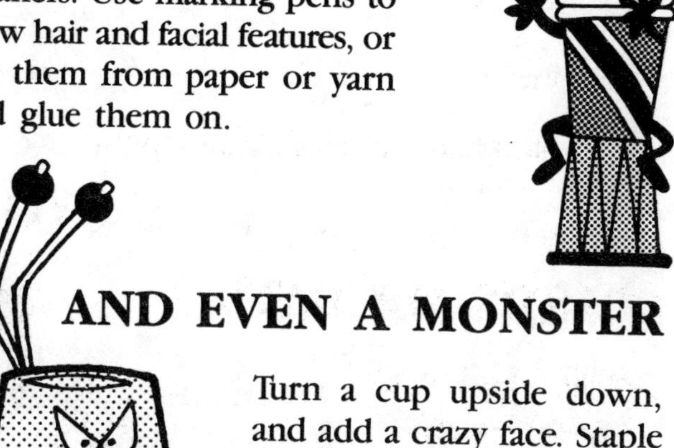

AND EVEN A MONSTER

Turn a cup upside down, and add a crazy face. Staple a cloth body to the edge of the cup. Glue on paper strip legs. Use brads, paper clips, toothpicks and other odds and ends to add antennae and special monstrous features. Give your monster a name like **Polly Polluter** or **Sir Lotsalitter**.

Recycled Box Puppets

Recycle used boxes by turning them into puppets. Use your puppets to stage a puppet show with an environmental theme.

WHAT YOU NEED

- ☐ used cardboard and paperboard boxes in assorted sizes (Consider candy, cereal, detergent, jewelry, match, shoe, and toothpaste boxes, as well as egg and milk cartons.)
- ☐ a pair of scissors
- ☐ rubber bands
- ☐ glue
- ☐ a roll of masking tape
- ☐ a stapler and staples
- ☐ a roll of shelf paper or some sheets of construction paper
- ☐ a pencil, felt-tipped marking pens, and/or crayons
- ☐ assorted odds and ends, such as brads, cotton balls, paper and fabric scraps, used paper cups and plates (rinsed and dried for reuse), feathers, paper clips, and pipe cleaners

YOU CAN MAKE A SHOE BOX ALLIGATOR

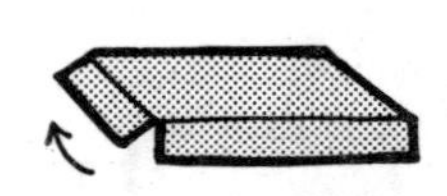

1. Cut the corners on one end of a shoe box lid and lift the flap.
2. Staple a rubber band to the top of the lid as shown.
3. Attach the lid to the box with masking tape.
4. Add eyes, teeth, and a tongue.
5. Put your fingers under the rubber band and move the lid up and down.

OR A CEREAL BOX KANGAROO

1. Cut open one end of a cereal box.
2. Turn the box so that the open end is at the bottom.
3. Glue on a paper cup for the nose and a small box for the pouch.
4. Cut ears, legs, and a tail from paper or cardboard and glue them on.
5. Use a tiny box to make a baby kangaroo and put it in the pouch.

Recycled Box Puppets
(continued)

YOU CAN MAKE A DETERGENT BOX GIANT

1. Cut off one end of a detergent box.

2. Cover the box with shelf or construction paper.

3. Use crayons, marking pens, or odds and ends to make your giant's face.

4. Cut a big body for your giant from paper or cloth and attach it to the front of the box.

5. To hold the giant puppet, put your hand inside the box behind the body.

OR A TOOTHPASTE BOX GIRAFFE

1. To make your giraffe's head, find two toothpaste boxes that are the same length.

2. Cut off one end of each box.

3. Stack the boxes with the open ends together.

4. Use a strip of masking tape to hinge the boxes together at the open ends.

5. Add eyes, ears, horns.

6. Cut, color, and attach a paper body to this head.

EARTHWORDS

The giraffe, in their queer, inimitable, vegetative gracefulness, as if it were not a herd of animals but a family of rare, long-stemmed, speckled gigantic flowers slowly advancing.

—Isak Dinesen [Karen Blixen]
Out of Africa

Paper Bag Vests

Turn a used paper bag into a brand new vest.

WHAT YOU NEED

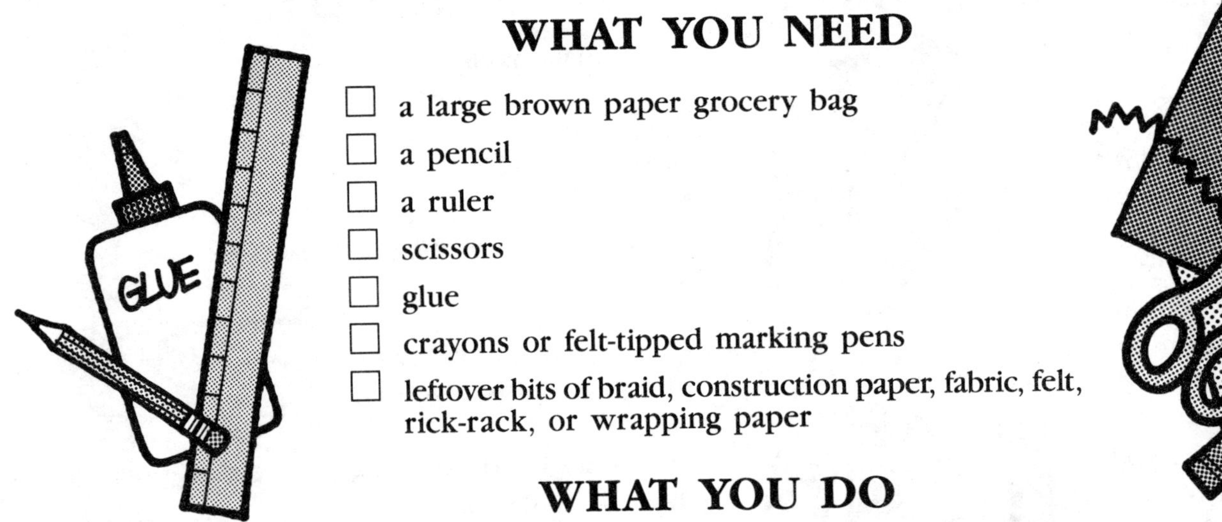

- ☐ a large brown paper grocery bag
- ☐ a pencil
- ☐ a ruler
- ☐ scissors
- ☐ glue
- ☐ crayons or felt-tipped marking pens
- ☐ leftover bits of braid, construction paper, fabric, felt, rick-rack, or wrapping paper

WHAT YOU DO

1. Using the pencil and ruler, draw a straight line up the middle of the front or back of the grocery bag.

2. With the scissors, carefully cut along this line.

3. On the bottom of the bag, draw a circle large enough to go around your neck. (Keep the circle small. Don't let it touch the edges of the bottom.)

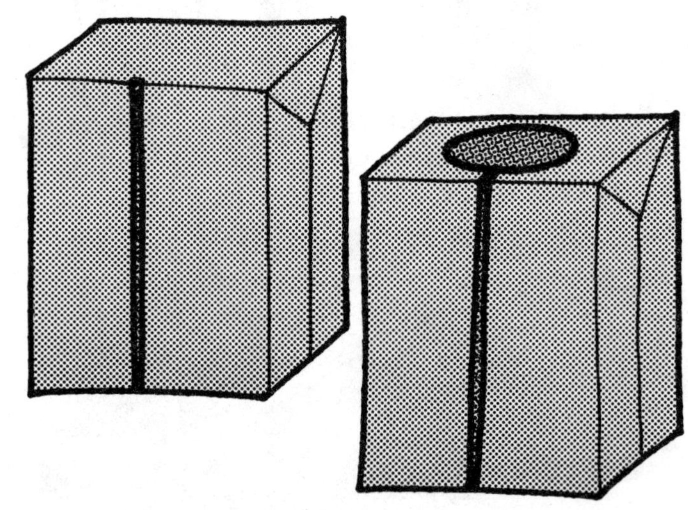

Paper Bag Vests
(continued)

4. Cut out this circle.

5. Cut rectangular holes in the side panels for your arms.

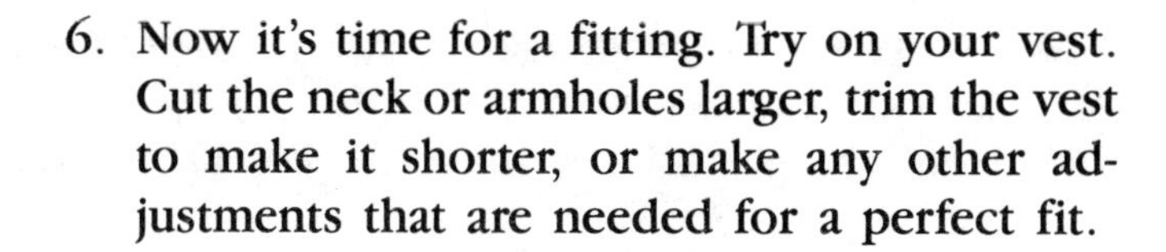

6. Now it's time for a fitting. Try on your vest. Cut the neck or armholes larger, trim the vest to make it shorter, or make any other adjustments that are needed for a perfect fit.

7. Fringe or scallop the bottom of the vest.

8. If there is printing on your grocery bag, glue a piece of colored paper over it.

9. Decorate your vest by drawing on it with crayons or felt-tipped pens and/or by gluing on bits of braid, felt, or rick-rack.

10. Write a slogan on your vest which will remind people to reuse and/or recycle.

EARTHWORDS

Where the forest murmurs there is music: ancient, everlasting. Go to the winter woods: listen there, look, watch, and "the dead months" will give you a subtler secret than any you have yet found in the forest.
—Fiona Macleod
Where the Forest Murmurs

Earth-Saver Award

Make an award to present to a family member, a friend, or a business in your community that has done something kind for the earth. For example, your award might recognize the recipient for being a *Careful Consumer*, a *Habitat Helper*, a *Responsible Recycler*, or a *Stream Saver*.

Design a Poster

Design a poster telling others not to litter. For example, your poster might warn kids *not* to release helium-filled balloons. Balloons that come down on water are life-threatening to birds and fish when swallowed. Balloons with metallic surfaces pose another threat to the environment. If they come in contact with exposed electrical wires, they can cause sparks, shorts, and power outages.

The Sad Saga of Benjamin Mark

A kid named Benjamin Mark
Lost his Mylar balloon in the park.
It sailed higher and higher—
Got stuck on a wire—
Now Ben is lost in the dark.

EARTHWORDS

We abuse land because we regard it as a commodity belonging to us. When we see land as a community to which we belong, we may begin to use it with love and respect.

—Aldo Leopold
A Sand County Almanac

A Shoe Box Diorama

Recycle a shoe box into a diorama to show at least one of the ways in which we are littering the earth.

WHAT YOU NEED

- ☐ a shoe box
- ☐ a brown paper grocery sack
- ☐ pieces of colored construction paper
- ☐ scissors
- ☐ glue
- ☐ scraps of fabric, foil, newspaper, metal, and/or plastic

WHAT YOU DO

1. Use the brown paper from the grocery sack to cover the outside of your shoe box.

2. Lay your covered shoe box on its side.

3. Select a scene. For example, your scene might be a city street, a beach, a park, a playground, a lake, a river, a public dump, or a landfill.

4. Using construction paper cutouts and scrap materials, create this scene inside the shoe box. To give your scene depth, include a background, a middle ground, and a foreground.

5. Share your finished diorama by displaying it in your home or classroom, or at your school or public library.

WHERE TO WRITE & GLOSSARY

How to Request Information

Earth-friendly organizations are working in your community, state, and country, and in other countries throughout the world to encourage recycling, save endangered plants and animals, protect habitats, and clean up the environment. The names and addresses of some of these organizations are listed on pages 173–178. On request, many of them will send you information about what they are doing and how you can help. When you write to these organizations, remember to do these four things.

1. Include your name and a complete return address so that the organizations to which you are writing will know where to send the information you have requested.

2. Enclose *two* first-class stamps with your request. Many of these organizations have limited budgets. The stamps you enclose will be used as postage for the information they send to you.

3. Be patient. Many of these organizations are understaffed, and the members of their staffs are overworked. Expect to wait three to six weeks for a reply.

4. Make the most of any information you receive. Share it with classmates, family members, and friends.

Energy Resources

Alliance to Save Energy
1725 K Street, N.W., Suite 914
Washington, D.C. 20006

American Council for an Energy-Efficient Economy
1001 Connecticut Avenue, N.W., Suite 535
Washington, D.C. 20036

California Energy Company
601 California Street, Suite 900
San Francisco, California 94108

Conservation and Renewable Energy Inquiry and Referral Service
P.O. Box 8900
Silver Spring, Maryland 20907

Geothermal Education Office
664 Hilary Drive
Tiburon, California 94920

National Appropriate Technology Assistance Service
P.O. Box 2525
Butte, Montana 59702-2525

National Energy Foundation (NEF)
Resources for Education, National Office
5160 Wiley Post Way, Suite 200
Salt Lake City, Utah 84116
(***Note:*** NEF materials were developed for educational purposes and are available to teachers only.)

Renew America
1400 16th Street, N.W., Suite 710
Washington, D.C. 20036

Rocky Mountain Institute
1739 Snowmass Creek Road
Snowmass, Colorado 81654

Solar Energy Research Institute
1617 Cole Boulevard
Golden, Colorado 80401

Environmental Resources

Acid Rain Foundation, Inc.
1410 Varsity Drive
Raleigh, North Carolina 27606

America the Beautiful Fund
219 Shoreham Building
Washington, D.C. 20005

American Forestry Association
P.O. Box 2000
Washington, D.C. 20010

American Water Works Association
6666 W. Quincy
Denver, Colorado 80235

Audubon Adventures
Route 4
Sharon, Connecticut 06069

Californians Against Waste
909 12th Street, Suite 201
Sacramento, California 95814

Citizens Clearinghouse for Hazardous Waste
P.O. Box 926
Arlington, Virginia 22216

Citizens for a Better Environment
924 Market Street, Suite 505
San Francisco, California 94102

Clean Water Action
317 Pennsylvania Avenue, S.E.
Washington, D.C. 20003

Earth Birthday Project
183 Pinehurst, #34
New York, New York 10033

Environmental Action Coalition
625 Broadway, 2nd Floor
New York, New York 10012

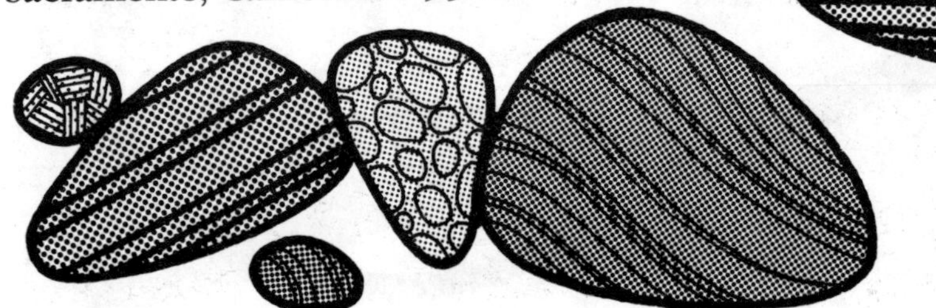

Environmental Resources
(continued)

Environmental Action Foundation
1525 New Hampshire Avenue, N.W.
Washington, D.C. 20036

Environmental Defense Fund
257 Park Avenue, South
New York, New York 10010

Environmental Hazards Management Institute (EHMI)
P.O. Box 932
10 Newmarket Road
Durham, New Hampshire 03824

Friends of the Earth
218 D Street, S.E.
Washington, D.C. 20003

Global ReLeaf
The American Forestry Association
P.O. Box 2000
Washington, D.C. 20013

Greenpeace
1611 Connecticut Avenue, N.W.
Washington, D.C. 20009

The Institute for Earth Education
P.O. Box 288
Warrenville, Illinois 60555

Institute for Environmental Education
32000 Chagrin Boulevard
Cleveland, Ohio 44124

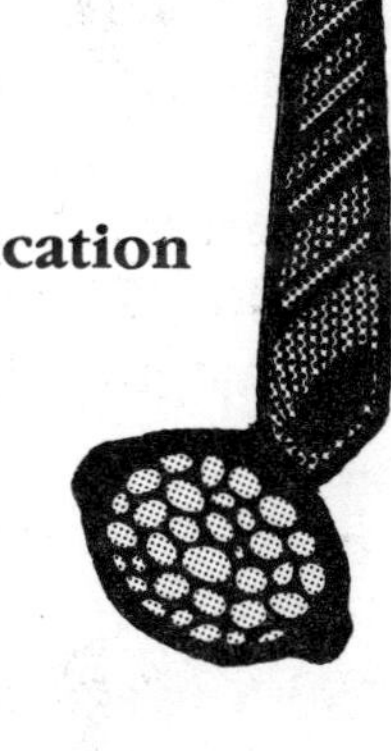

Institute for Local Self-Reliance
2425 18th Street, N.W.
Washington, D.C. 20009

The Izaak Walton League of America
1401 Wilson Boulevard, Level B
Arlington, Virginia 22209

Keep America Beautiful
9 West Broad Street
Stamford, Connecticut 06892

Environmental Resources
(continued)

Long Branch Environmental Education Center
Route 2, Box 132
Leichester, New York 28748

National Arbor Day Foundation
100 Arbor Avenue
Nebraska City, Nebraska 68410

National Recycling Coalition
1101 30th Street, N.W.
Suite 305
Washington, D.C. 20007

The Natural Resources Defense Council
40 West 20th Street
New York, New York 10011

The Nature Conservancy
1815 North Lynn Street
Arlington, Virginia 22209

The New Alchemy Institute
237 Hatchville Road
East Falmouth, Massachusetts 02536

Sierra Club
730 Polk Street
San Francisco, California 94109

Trees for Life
1103 Jefferson Street
Wichita, Kansas 67203

U.S. Environmental Protection Agency (EPA)
401 M Street, S.W.
A 108
Washington, D.C. 20460

Western Regional Environmental Education Council
2820 Echo Way
Sacramento, California 95821

World Federalist Association
United Nations Office
777 United Nations Plaza
New York, New York 10017

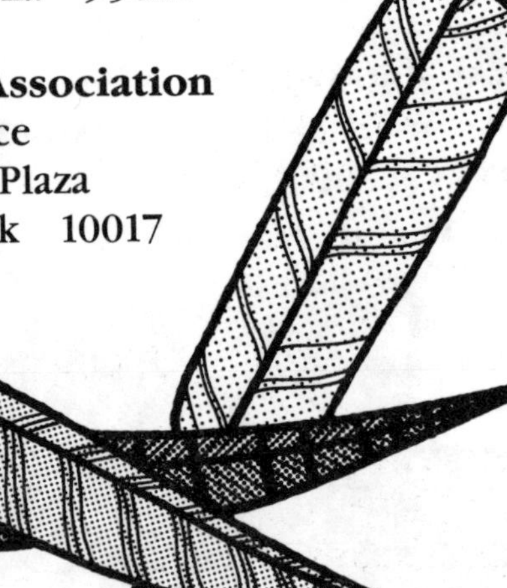

Wildlife Resources

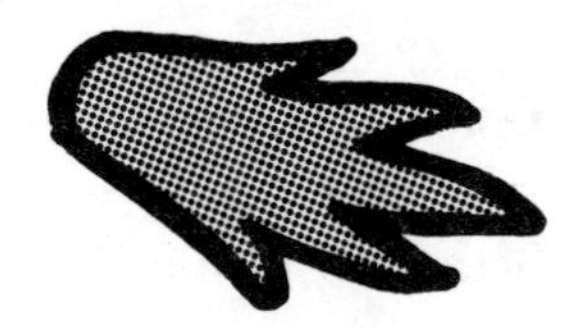

Animal Welfare Institute
P.O. Box 3650
Washington, D.C. 20007

Center for Action on Endangered Species
175 West Main Street
Ayer, Massachusetts 01432

Center for Marine Conservation
1725 DeSalles Street, N.W.
Suite 500
Washington, D.C. 20036

Defenders of Wildlife
1244 19th Street, N.W.
Washington, D.C. 20036

Earth Island Institute
Save the Dolphins
300 Broadway, Suite 28
San Francisco, California 94133-3312

Greenpeace
1436 U Street, N.W.
Washington, D.C. 20009

The Humane Society of the United States
2100 L Street, N.W.
Washington, D.C. 20037

The International Crane Foundation
E-11376 Shady Lane Road
Baraboo, Wisconsin 53913

National Association for the Advancement of Humane Education
67 Salem Road
East Haddam, Connecticut 06423

National Audubon Society
950 Third Avenue
New York, New York 10022

National Wildlife Federation
1412 16th Street, N.W.
Washington, D.C. 20036

U.S. Fish and Wildlife Service
Publications Unit
130 Arlington Square Building
18th and C Streets, N.W.
Washington, D.C. 20240

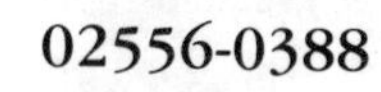

Whale Adoption Project
P.O. Box 388
North Falmouth, Massachusetts 02556-0388

World Wildlife Fund
P.O. Box 96220
Washington, D.C. 20077-7787

Canadian Resources

Canadian Nature Federation
453 Sussex Drive
Ottawa, Ontario K1N 6Z4

Ecology Action Centre
1657 Barrington Street, Suite 520
Halifax, Nova Scotia B3J 2A1

Energy Probe
100 College Street
Toronto, Ontario M5G 1L5

Friends of the Earth
53 Queen Street, Room 16
Ottawa, Ontario K1P 5C5

Greenpeace
427 Bloor Street West
Toronto, Ontario M5S 1X7

Society Promoting Environmental Conservation
2150 Maple Street
Vancouver, British Columbia V6J 3T3

World-Wide Fund for Nature
60 St. Clair Avenue East, Suite 201
Toronto, Ontario M4T 1N5

Glossary

acid a water-soluble chemical compound that tastes sour or bitter, irritates skin and eyes, and reddens litmus paper (page 90)

acid rain a harmful type of precipitation which occurs when airborne chemicals like sulfur dioxide and nitrogen dioxide become dissolved in rainwater (page 90)

air pollutant anything that makes air dirty or impure (page 102)

aqueduct a channel built to carry water from where it is to where it is needed (page 72)

battery a container holding materials that produce electrical energy by chemical reaction (page 61)

biodegradable capable of being broken down by the action of bacteria and other microorganisms into products that will not harm the environment

carbon dioxide a colorless, odorless, tasteless gas that is produced when animals exhale and when fuels burn and is used by plants to make food (pages 106, 107)

carnivores animals that eat other animals and are nourished by the plants and smaller creatures these animals have eaten (page 107)

chlorofluorocarbons (CFCs) chemicals that are present in aerosol mixtures, are released when polystyrene is being made, and are believed to thin the ozone layer (page 44)

chlorophyll a green substance that enables the leaves of plants to use solar energy, carbon dioxide, water, and organic nutrients from the soil to make the sugars and starches they use as food (page 106)

composting the process of turning organic wastes into a nutrient-rich mixture that can be used to condition soil and feed plants (page 112)

decompose to rot or decay; to break down into simpler parts or elements

decomposers mushrooms, insects, worms, and other organisms that feed on decaying plant and animal matter and break it down into a form that can be used as nutrients by plants (page 107)

deforestation the complete destruction and total clearing of all forests within a region (page 108)

deposit the amount of money that is charged when a product is purchased in a refundable container to encourage return of the container (page 52)

detergent a cleansing agent; a liquid or powder used to wash dishes or clothes (page 88)

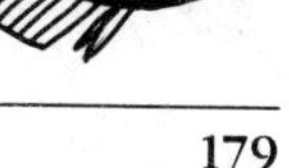

Glossary
(continued)

ecology the science that studies the ways in which organisms and their environments are interrelated

endangered animals specific kinds of animals whose total population is becoming steadily smaller, or decreasing, so that they are in danger of becoming extinct (pages 118, 120–121)

energy the capacity to do work or the ability to make things move (page 56)

environment all of the natural and living things with which we are surrounded; the climate and conditions in which any organism lives

extinct no longer existing in living or active form (page 118)

fossil fuels certain natural substances, such as coal, oil, and natural gas, which were created deep within the earth millions of years ago by the decomposing remains of plants and animals and can be burned to release energy (pages 62, 94)

glass a very hard substance that is made by melting sand with certain chemicals, breaks easily, and can usually be seen through (page 52)

global warming an overall increase in the earth's temperature which may be caused by reduced numbers of trees and increased levels of carbon dioxide (page 109)

groundwater water that lies under the ground in natural reservoirs, such as springs and wells (page 72)

habitat the place where any plant or animal naturally lives and grows; a place that provides food, water, space to live, and shelter for an interdependent community of living things, including both plants and animals (page 116)

herbivores animals that rely on plants and plant parts for their nourishment (page 107)

incinerate to burn (page 24)

kilowatt a unit of measure for electricity which equals one thousand watts (page 60)

kilowatt-hour the amount of energy present in one kilowatt of electricity supplied for one hour of time; the amount of electrical energy needed to keep a standard 100-watt bulb burning for 10 hours (page 60)

kinetic energy the active form of energy that is found in heat, light, sound, and motion (page 56)

landfill an enormous pit where trash is buried under shallow layers of dirt (page 24)

litter any unneeded item that has been carelessly discarded instead of being disposed of properly (page 20)

Glossary (continued)

mains large distribution pipes that carry water from the plants where it is treated to the homes, schools, and businesses where it is used (page 75)

nonbiodegradable incapable of being broken down naturally into substances that will not harm the environment

nonrenewable resource a resource that cannot be replaced, replenished, or renewed by natural processes or by human planning and practices (pages 43, 94)

oil slick a floating film that is formed atop a body of water by oil that has been spilled into it (page 96)

organic related to living things; made by or gotten from plants or animals (page 55)

oxygen a colorless, odorless, tasteless gas that is produced by plants and needed by animals and people (pages 106, 107)

ozone layer a layer of oxygen formed naturally, high above the earth, which acts as a screen to protect plants and animals from the sun's ultraviolet rays (page 44)

pesticide a poison that is used to kill pests, such as insects, rodents, or weeds (page 99)

pesticide residue the small amount of poison that may remain on plants long after they have been sprayed with pesticide and may be eaten by wild creatures, farm animals, and/or people (page 99)

petroleum an oily, flammable liquid that ranges in color from clear to black and is found in the rocks which form the earth's crust (page 94)

phosphates chemicals added to detergents to increase the amount of suds and make these suds last longer (page 88)

plastic the name given to a large group of substances made chemically from such materials as coal or oil mixed with water and limestone (page 43)

pollutant any substance that can make air, land, or water dirty or impure

polystyrene the main ingredient in foam plastics (page 44)

potential energy the stored form of energy available in resources such as coal, gasoline, oil, and water held behind a dam (page 56)

rain forests tropical woodlands that grow in the hot, humid areas of the earth, near its equator, and receive an annual rainfall of at least 100 inches (page 114)

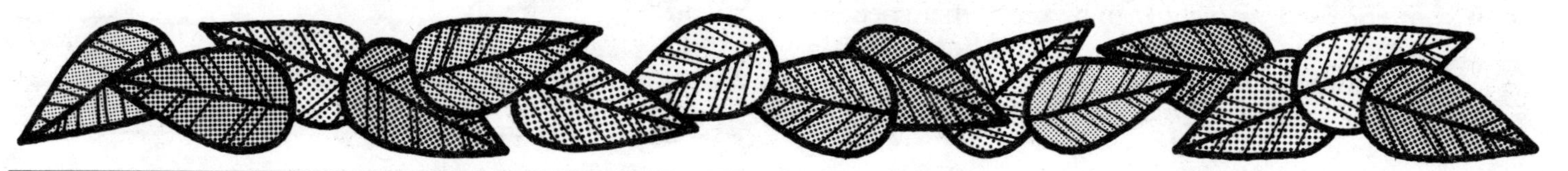

Glossary
(continued)

recycle to process and treat discarded materials so that they can be used again (pages 16, 18)

redeem to buy back (page 52)

redemption value the amount of money that will be paid when a refundable container is redeemed (page 52)

reservoirs natural or man-made structures in which water can be stored for future use (pages 72, 75)

resources substances that support life and fulfill human needs, including air, land, water, minerals, fossil fuels, forests, and sunlight

solar energy energy produced by the sun (page 63)

topsoil the rich organic layer of dirt from which plants get the nutrients they need but which may be washed away by rainwater if left bare and unprotected (page 109)

toxic substance a chemical or mixture of chemicals whose manufacture, distribution, use, or disposal may present an unreasonable risk to the health of a person and/or the environment (page 100)

ultraviolet rays a form of radiation that is present in sunlight but is too short to be seen, that speeds healing and aids the formation of vitamins but is harmful in large amounts (page 44)

water cycle the continuous natural process by which water evaporates from bodies of water, collects in the atmosphere as vapor, condenses in clouds, falls to the ground as rain, and evaporates once again (page 70)

wetlands low-lying areas—including bogs, deltas, lakes, marshes, ponds, or swamps—that are saturated with moisture and provide food-rich habitats for a wide variety of animals and plants (page 113)

wind farm acres of flat or gently rolling land in naturally windy areas on which people build rows of windmills instead of planting rows of crops (page 62)

windmill a mechanical device that consists of blades attached to a central pole and uses wind energy to generate electricity (page 62)

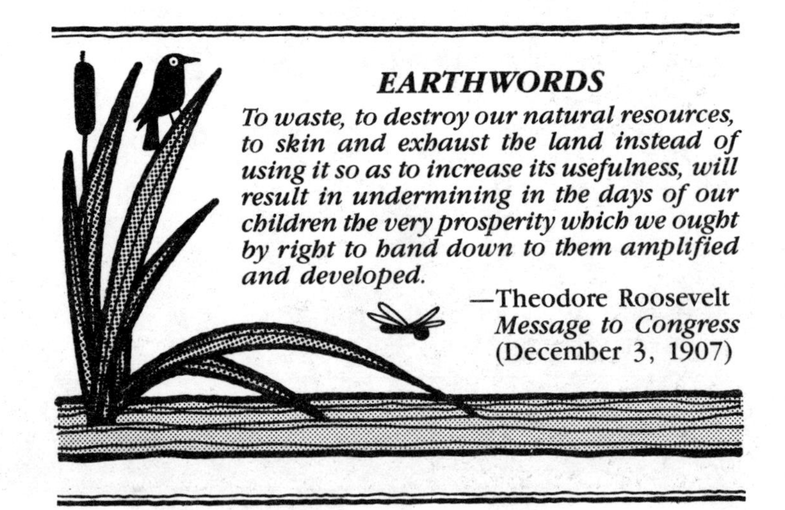

EARTHWORDS

To waste, to destroy our natural resources, to skin and exhaust the land instead of using it so as to increase its usefulness, will result in undermining in the days of our children the very prosperity which we ought by right to hand down to them amplified and developed.

—Theodore Roosevelt
Message to Congress
(December 3, 1907)

Closing Thoughts

Earth Thoughts

How long can our earth survive?
Nobody knows.
People don't realize what trouble our planet is in.
They only hear the birds chirping happily
and smell the roses blooming.
They only want to believe the good things in life—
not think about the earth's problems.
Maybe if everyone realized what troubles the earth
faces—polluted water and air, destruction of
our rain forests, litter everywhere—
Maybe if everyone did his share to help
do something about these problems,
Then people could go back to thoughts of
the sound of birds chirping
And the smell of roses blooming.

—Mike Schwartz
Grade 6

Closing Thoughts
(continued)

Earth

Earth is like a bird
singing cheerfully,
like a red ribbon curled,
a fish wanting to go upstream,
keeping the environment clean
like the earth was just born.

Hear Mother Nature and
Mother Earth say, "Don't wreck
this place that we made!
Don't pollute the water
and land. Would you want
someone to wreck something
you had made?"

The air is the place where
outer space is, a place where
we want to live and to see
what's around planets, where
we want to grow life.

Sky is where our sweet air
is stored. If it was all plugged up,
we would have problems breathing
and all the birds would die—
no more beautiful songs,
no more plants would grow,
and we would not see the sun
or rainbows anymore.

—Brian Haeberle
Grade 3

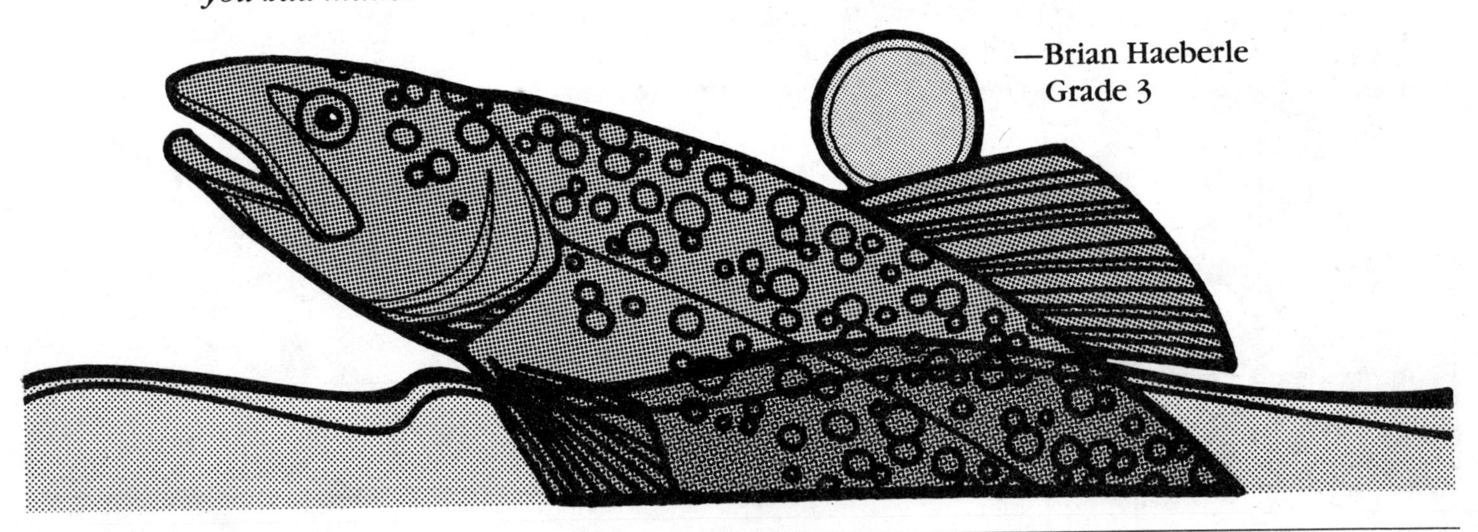